OFFICIAL SQA PAST PAPERS WITH ANSWERS

HIGHER

PHYSICS
2008-2012

2008 EXAM – page 3
2009 EXAM – page 29
2010 EXAM – page 53
2011 EXAM – page 77
2012 EXAM – page 103
ANSWER SECTION – page 131

First exam published in 2008.

Published by Bright Red Publishing Ltd, 6 Stafford Street, Edinburgh EH3 7AU

tel: 0131 220 5804 fax: 0131 220 6710 info@brightredpublishing.co.uk www.brightredpublishing.co.uk

ISBN 978-1-84948-297-4

A CIP Catalogue record for this book is available from the British Library.

Bright Red Publishing is grateful to the copyright holders, as credited on the final page of the Question Section, for permission to use their material. Every effort has been made to trace the copyright holders and to obtain their permission for the use of copyright material. Bright Red Publishing will be happy to receive information allowing us to rectify any error or omission in future editions.

[BLANK PAGE]

X069/301

NATIONAL
QUALIFICATIONS
2008

FRIDAY, 23 MAY
1.00 PM – 3.30 PM

PHYSICS
HIGHER

Read Carefully

Reference may be made to the Physics Data Booklet.

1 All questions should be attempted.

Section A (questions 1 to 20)

2 Check that the answer sheet is for Physics Higher (Section A).

3 For this section of the examination you must use an **HB pencil** and, where necessary, an eraser.

4 Check that the answer sheet you have been given has **your name**, **date of birth**, **SCN** (Scottish Candidate Number) and **Centre Name** printed on it.
   Do not change any of these details.

5 If any of this information is wrong, tell the Invigilator immediately.

6 If this information is correct, **print** your name and seat number in the boxes provided.

7 There is **only one correct** answer to each question.

8 Any rough working should be done on the question paper or the rough working sheet, **not** on your answer sheet.

9 At the end of the exam, put the **answer sheet for Section A inside the front cover of your answer book**.

10 Instructions as to how to record your answers to questions 1–20 are given on page three.

Section B (questions 21 to 30)

11 Answer the questions numbered 21 to 30 in the answer book provided.

12 **All answers must be written clearly and legibly in ink**.

13 Fill in the details on the front of the answer book.

14 Enter the question number clearly in the margin of the answer book beside each of your answers to questions 21 to 30.

15 Care should be taken to give an appropriate number of significant figures in the final answers to calculations.

16 Where additional paper, eg square ruled paper, is used, write your name and SCN (Scottish Candidate Number) on it and place it inside the front cover of your answer booklet.

DATA SHEET
COMMON PHYSICAL QUANTITIES

Quantity	Symbol	Value	Quantity	Symbol	Value
Speed of light in vacuum	c	3.00×10^8 m s^{-1}	Mass of electron	m_e	9.11×10^{-31} kg
Magnitude of the charge on an electron	e	1.60×10^{-19} C	Mass of neutron	m_n	1.675×10^{-27} kg
Gravitational acceleration on Earth	g	9.8 m s^{-2}	Mass of proton	m_p	1.673×10^{-27} kg
Planck's constant	h	6.63×10^{-34} J s			

REFRACTIVE INDICES

The refractive indices refer to sodium light of wavelength 589 nm and to substances at a temperature of 273 K.

Substance	Refractive index	Substance	Refractive index
Diamond	2·42	Water	1·33
Crown glass	1·50	Air	1·00

SPECTRAL LINES

Element	Wavelength/nm	Colour	Element	Wavelength/nm	Colour
Hydrogen	656	Red	Cadmium	644	Red
	486	Blue-green		509	Green
	434	Blue-violet		480	Blue
	410	Violet		*Lasers*	
	397	Ultraviolet			
	389	Ultraviolet	Element	Wavelength/nm	Colour
Sodium	589	Yellow	Carbon dioxide	9550 / 10590	Infrared
			Helium-neon	633	Red

PROPERTIES OF SELECTED MATERIALS

Substance	Density/ kg m^{-3}	Melting Point/ K	Boiling Point/ K
Aluminium	2.70×10^3	933	2623
Copper	8.96×10^3	1357	2853
Ice	9.20×10^2	273	. . . .
Sea Water	1.02×10^3	264	377
Water	1.00×10^3	273	373
Air	1·29	. . . .	. . . .
Hydrogen	9.0×10^{-2}	14	20

The gas densities refer to a temperature of 273 K and a pressure of 1.01×10^5 Pa.

SECTION A

For questions 1 to 20 in this section of the paper the answer to each question is either A, B, C, D or E. Decide what your answer is, then, using your pencil, put a horizontal line in the space provided—see the example below.

EXAMPLE

The energy unit measured by the electricity meter in your home is the

  A   kilowatt-hour

  B   ampere

  C   watt

  D   coulomb

  E   volt.

The correct answer is **A**—kilowatt-hour. The answer **A** has been clearly marked in **pencil** with a horizontal line (see below).

Changing an answer

If you decide to change your answer, carefully erase your first answer and, using your pencil, fill in the answer you want. The answer below has been changed to **E**.

[Turn over

SECTION A

Answer questions 1–20 on the answer sheet.

1. Which row in the table is correct?

	Scalar	Vector
A	distance	work
B	weight	acceleration
C	velocity	displacement
D	mass	momentum
E	speed	time

2. A javelin is thrown at $60°$ to the horizontal with a speed of $20\ \text{m s}^{-1}$.

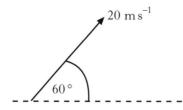

The javelin is in flight for $3.5\ \text{s}$.
Air resistance is negligible.
The horizontal distance the javelin travels is

A $35.0\ \text{m}$

B $60.6\ \text{m}$

C $70.0\ \text{m}$

D $121\ \text{m}$

E $140\ \text{m}$.

3. Two boxes on a frictionless horizontal surface are joined together by a string. A constant horizontal force of 12 N is applied as shown.

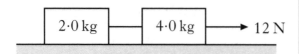

The tension in the string joining the two boxes is

A $2.0\ \text{N}$

B $4.0\ \text{N}$

C $6.0\ \text{N}$

D $8.0\ \text{N}$

E $12\ \text{N}$.

4. The total mass of a motorcycle and rider is $250\ \text{kg}$. During braking, they are brought to rest from a speed of $16.0\ \text{m s}^{-1}$ in a time of $10.0\ \text{s}$.

The maximum energy which could be converted to heat in the brakes is

A $2000\ \text{J}$

B $4000\ \text{J}$

C $32\,000\ \text{J}$

D $40\,000\ \text{J}$

E $64\,000\ \text{J}$.

5. A shell of mass $5.0\ \text{kg}$ is travelling horizontally with a speed of $200\ \text{m s}^{-1}$. It explodes into two parts. One part of mass $3.0\ \text{kg}$ continues in the original direction with a speed of $100\ \text{m s}^{-1}$.

The other part also continues in this same direction. Its speed is

A $150\ \text{m s}^{-1}$

B $200\ \text{m s}^{-1}$

C $300\ \text{m s}^{-1}$

D $350\ \text{m s}^{-1}$

E $700\ \text{m s}^{-1}$.

6. The graph shows the force which acts on an object over a time interval of 8 seconds.

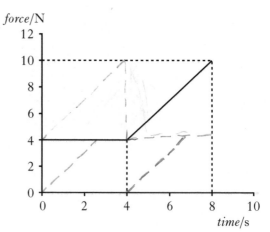

The momentum gained by the object during this 8 seconds is

A $12\ \text{kg m s}^{-1}$

B $32\ \text{kg m s}^{-1}$

C $44\ \text{kg m s}^{-1}$

D $52\ \text{kg m s}^{-1}$

E $72\ \text{kg m s}^{-1}$.

7. One pascal is equivalent to

A $1 \, \text{N m}$

B $1 \, \text{N m}^2$

C $1 \, \text{N m}^3$

D $1 \, \text{N m}^{-2}$

E $1 \, \text{N m}^{-3}$.

8. An electron is accelerated from rest through a potential difference of $2 \cdot 0 \, \text{kV}$.

The kinetic energy gained by the electron is

A $8 \cdot 0 \times 10^{-23} \, \text{J}$

B $8 \cdot 0 \times 10^{-20} \, \text{J}$

C $3 \cdot 2 \times 10^{-19} \, \text{J}$

D $1 \cdot 6 \times 10^{-16} \, \text{J}$

E $3 \cdot 2 \times 10^{-16} \, \text{J}$.

9. The e.m.f. of a battery is

A   the total energy supplied by the battery

B   the voltage lost due to the internal resistance of the battery

C   the total charge which passes through the battery

D   the number of coulombs of charge passing through the battery per second

E   the energy supplied to each coulomb of charge passing through the battery.

10. The diagram shows the trace on an oscilloscope when an alternating voltage is applied to its input.

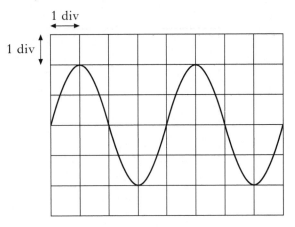

The timebase is set at $5 \, \text{ms/div}$ and the Y-gain is set at $10 \, \text{V/div}$.

Which row in the table gives the peak voltage and the frequency of the signal?

	Peak voltage/V	Frequency/Hz
A	7·1	20
B	14	50
C	20	20
D	20	50
E	40	50

[Turn over

11. A resistor is connected to an a.c. supply as shown.

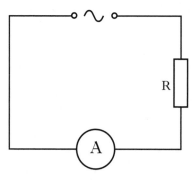

a.c. ammeter

The supply has a constant peak voltage, but its frequency can be varied.

The frequency is steadily increased from 50 Hz to 5000 Hz.

The reading on the a.c. ammeter

A   remains constant

B   decreases steadily

C   increases steadily

D   increases then decreases

E   decreases then increases.

12. An ideal op-amp is connected as shown.

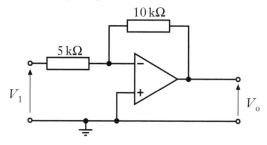

The graph shows how the input voltage, V_1, varies with time.

Which graph shows how the output voltage, V_o, varies with time?

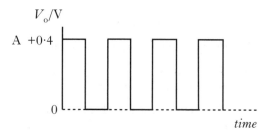

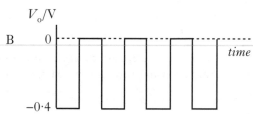

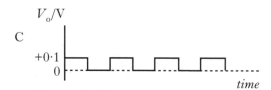

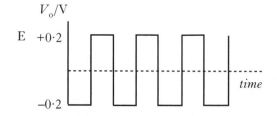

13. Which of the following proves that light is transmitted as waves?

A   Light has a high velocity.

B   Light can be reflected.

C   Light irradiance reduces with distance.

D   Light can be refracted.

E   Light can produce interference patterns.

14. A source of microwaves of wavelength λ is placed behind two slits, R and S.

A microwave detector records a maximum when it is placed at P.

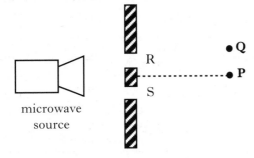

microwave
source

The detector is moved and the **next** maximum is recorded at Q.

The path difference $(SQ - RQ)$ is

A 0

B $\dfrac{\lambda}{2}$

C λ

D $\dfrac{3\lambda}{2}$

E 2λ.

15. A student makes five separate measurements of the diameter of a lens.

These measurements are shown in the table.

Diameter of lens/mm	22·5	22·6	22·4	22·6	22·9

The approximate random uncertainty in the mean value of the diameter is

A   0·1 mm

B   0·2 mm

C   0·3 mm

D   0·4 mm

E   0·5 mm.

16. The value of the absolute refractive index of diamond is 2·42.

The critical angle for diamond is

A   0·413°

B   24·4°

C   42·0°

D   65·6°

E   90·0°.

[Turn over

17. Part of the energy level diagram for an atom is shown.

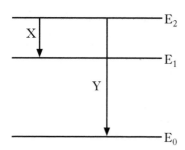

X and Y represent two possible electron transitions.
Which of the following statements is/are correct?

I Transition Y produces photons of higher frequency than transition X.

II Transition X produces photons of longer wavelength than transition Y.

III When an electron is in the energy level E_0, the atom is ionised.

A I only

B I and II only

C I and III only

D II and III only

E I, II and III

18. The letters **X**, **Y** and **Z** represent three missing words from the following passage.

Materials can be divided into three broad categories according to their electrical resistance.

.............**X**............. *have a very high resistance.*

.............**Y**............. *have a high resistance in their pure form but when small amounts of certain impurities are added, the resistance decreases.*

.............**Z**............. *have a low resistance.*

Which row in the table shows the missing words?

	X	**Y**	**Z**
A	conductors	insulators	semi-conductors
B	semi-conductors	insulators	conductors
C	insulators	semi-conductors	conductors
D	conductors	semi-conductors	insulators
E	insulators	conductors	semi-conductors

19. Compared with a proton, an alpha particle has

A twice the mass and twice the charge

B twice the mass and the same charge

C four times the mass and twice the charge

D four times the mass and the same charge

E twice the mass and four times the charge.

20. For the nuclear decay shown, which row of the table gives the correct values of x, y and z?

$$^{214}_{x}\text{Pb} \longrightarrow {}^{y}_{83}\text{Bi} + {}^{0}_{z}\text{e}$$

	x	y	z
A	85	214	2
B	84	214	1
C	83	210	4
D	82	214	−1
E	82	210	−1

[Turn over for SECTION B on *Page ten*

SECTION B

Write your answers to questions 21 to 30 in the answer book.

Marks

21. To test the braking system of cars, a test track is set up as shown.

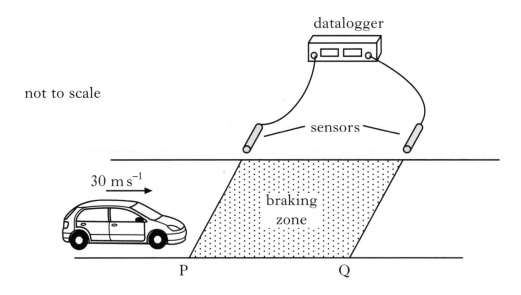

The sensors are connected to a datalogger which records the speed of a car at both P and Q.

A car is driven at a constant speed of 30 m s^{-1} until it reaches the start of the braking zone at P. The brakes are then applied.

(a) In one test, the datalogger records the speed at P as 30 m s^{-1} and the speed at Q as 12 m s^{-1}. The car slows down at a constant rate of $9 \cdot 0 \text{ m s}^{-2}$ between P and Q.

Calculate the length of the braking zone.   2

(b) The test is repeated. The same car is used but now with passengers in the car. The speed at P is again recorded as 30 m s^{-1}.

The same braking force is applied to the car as in part (a).

How does the speed of the car at Q compare with its speed at Q in part (a)? Justify your answer.   2

Marks

21. (continued)

(c) The brake lights of the car consist of a number of very bright LEDs.

An LED from the brake lights is forward biased by connecting it to a 12 V car battery as shown.

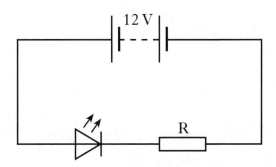

The battery has negligible internal resistance.

(i) Explain, in terms of charge carriers, how the LED emits light.   1

(ii) The LED is operating at its rated values of 5·0 V and 2·2 W.

Calculate the value of resistor R.   3

(8)

[Turn over

Marks

22. A crate of mass 40·0 kg is pulled up a slope using a rope.

    The slope is at an angle of 30° to the horizontal.

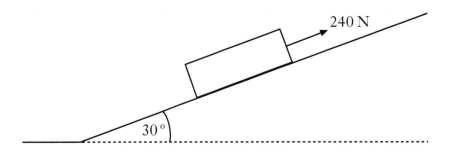

   A force of 240 N is applied to the crate parallel to the slope.

   The crate moves at a constant speed of 3·0 m s⁻¹.

   (*a*) (i) Calculate the component of the weight of the crate acting parallel to the slope. 2

   (ii) Calculate the frictional force acting on the crate. 2

   (*b*) As the crate is moving up the slope, the rope snaps.

   The graph shows how the velocity of the crate changes from the moment the rope snaps.

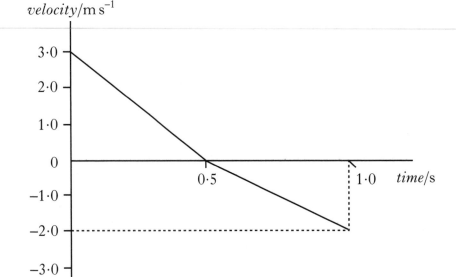

   (i) Describe the motion of the crate during the first 0·5 s after the rope snaps. 1

Marks

22. (b) (continued)

(ii) Copy the axes shown below and sketch the graph to show the acceleration of the crate between 0 and 1·0 s.

Appropriate numerical values are also required on the acceleration axis.

2

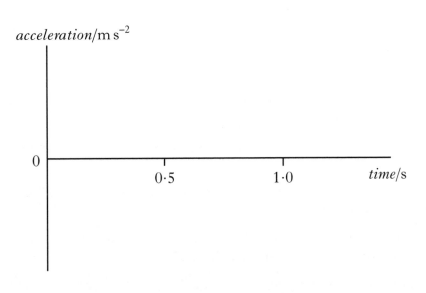

(iii) Explain, in terms of the forces acting on the crate, why the magnitude of the acceleration changes at 0·5 s.

2

(9)

[Turn over

Marks

23. A cylinder of compressed oxygen gas is in a laboratory.

(a) The oxygen inside the cylinder is at a pressure of $2\cdot82 \times 10^6\,\mathrm{Pa}$ and a temperature of $19\cdot0\,^\circ\mathrm{C}$.

The cylinder is now moved to a storage room where the temperature is $5\cdot0\,^\circ\mathrm{C}$.

   (i) Calculate the pressure of the oxygen inside the cylinder when its temperature is $5\cdot0\,^\circ\mathrm{C}$.   **2**

  (ii) What effect, if any, does this decrease in temperature have on the density of the oxygen in the cylinder?

Justify your answer.   **2**

(b) (i) The volume of oxygen inside the cylinder is $0\cdot030\,\mathrm{m}^3$.

The density of the oxygen inside the cylinder is $37\cdot6\,\mathrm{kg\,m}^{-3}$.

Calculate the mass of oxygen in the cylinder.   **2**

  (ii) The valve on the cylinder is opened slightly so that oxygen is gradually released.

The temperature of the oxygen inside the cylinder remains constant.

Explain, in terms of particles, why the pressure of the gas inside the cylinder decreases.   **1**

 (iii) After a period of time, the pressure of the oxygen inside the cylinder reaches a constant value of $1\cdot01 \times 10^5\,\mathrm{Pa}$. The valve remains open.

Explain why the pressure does not decrease below this value.   **1**

       (8)

Marks

24. Electrically heated gloves are used by skiers and climbers to provide extra warmth.

(*a*) Each glove has a heating element of resistance $3 \cdot 6\,\Omega$.

Two cells, each of e.m.f. $1 \cdot 5\,V$ and internal resistance $0 \cdot 20\,\Omega$, are used to operate the heating element.

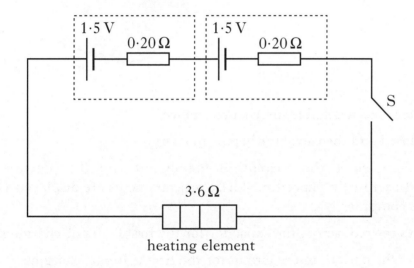

Switch S is closed.

    (i)   Determine the value of the total circuit resistance.   **1**

    (ii) Calculate the current in the heating element.   **2**

    (iii) Calculate the power output of the heating element.   **2**

(*b*) When in use, the internal resistance of each cell gradually increases.

What effect, if any, does this have on the power output of the heating element?

Justify your answer.   **2**

       (7)

[Turn over

25. (*a*) State what is meant by the term *capacitance*.   **1**

(*b*) An uncharged capacitor, C, is connected in a circuit as shown.

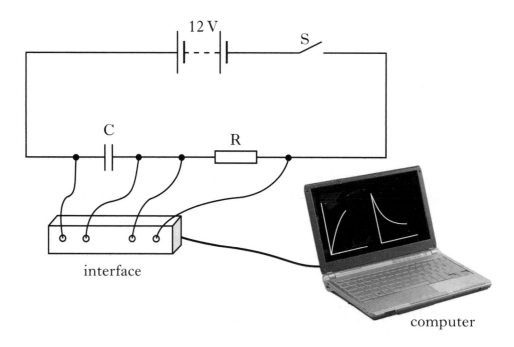

The 12 V battery has negligible internal resistance.

Switch S is closed and the capacitor begins to charge.

The interface measures the current in the circuit and the potential difference (p.d.) across the capacitor. These measurements are displayed as graphs on the computer.

Graph 1 shows the p.d. across the capacitor for the first 0·40 s of charging.

Graph 2 shows the current in the circuit for the first 0·40 s of charging.

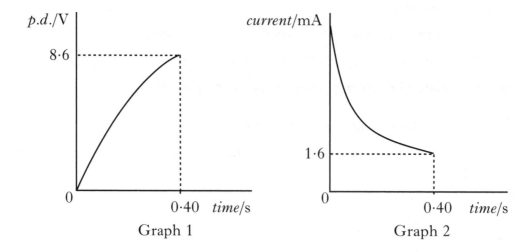

Marks

25. **(b)** **(continued)**

    (i)   Determine the p.d. **across resistor R** at 0·40 s.   **1**

    (ii) Calculate the resistance of R.   **2**

    (iii) The capacitor takes 2·2 seconds to charge fully.

          At that time it stores 10·8 mJ of energy.

          Calculate the capacitance of the capacitor.   **3**

  (c) The capacitor is now discharged.
A second, identical resistor is connected in the circuit as shown.

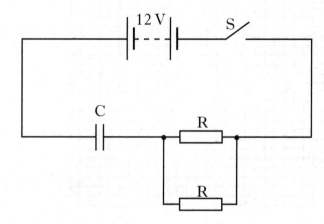

Switch S is closed.

Is the time taken for the capacitor to fully charge less than, equal to, or greater than the time taken to fully charge in part (b)?

Justify your answer.   **2**

                                                                (9)

[Turn over

Marks

26. The graph shows how the resistance of an LDR changes with the irradiance of light incident on it.

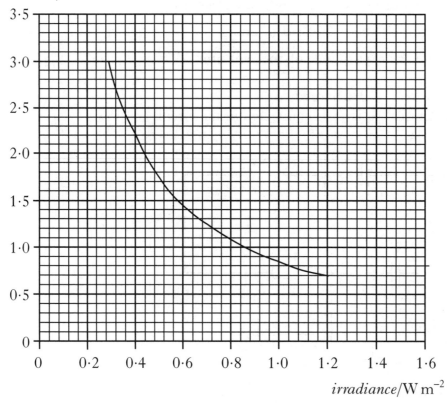

resistance/kΩ

irradiance/W m^{-2}

(a) The LDR is connected in the following bridge circuit.

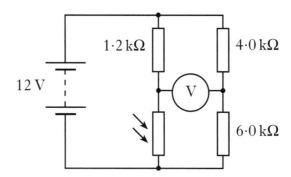

Determine the value of irradiance at which the bridge is balanced.

Show clearly how you arrive at your answer.

3

Marks

26.   (continued)

(*b*)   The LDR is now mounted on the outside of a car to monitor light level. It forms part of a circuit which provides an indication for the driver to switch on the headlamps.
The circuit is shown below.

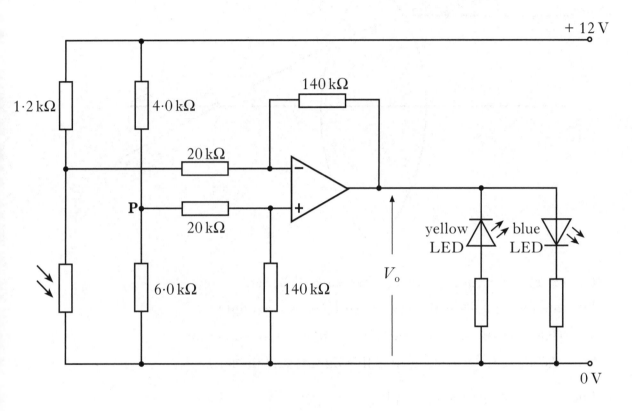

The LEDs inside the car indicate whether the headlamps should be on or off.

(i)   At a particular value of irradiance the resistance of the LDR is $2 \cdot 0 \, k\Omega$.

Show that the potential difference across the LDR in the circuit is $7 \cdot 5 \, V$.   1

(ii)   The potential at point P in the circuit is $7 \cdot 2 \, V$.

Calculate the output voltage, V_o, of the op-amp at this light level.   2

(iii)   Which LED(s) is/are lit at this value of output voltage?

Justify your answer.   2

(8)

[Turn over

Marks

27. (*a*) A ray of red light of frequency $4{\cdot}80 \times 10^{14}$ Hz is incident on a glass lens as shown.

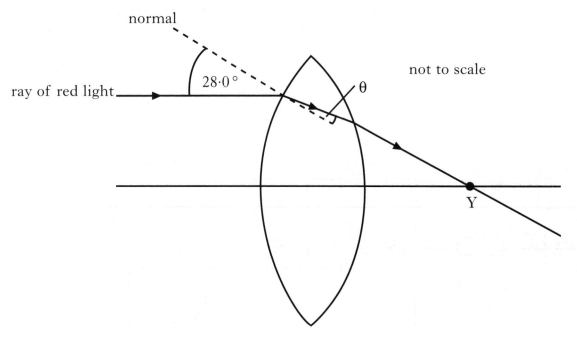

The ray passes through point Y after leaving the lens.

The refractive index of the glass is 1·61 for this red light.

    (i)   Calculate the value of the angle θ shown in the diagram.   **2**

   (ii)   Calculate the wavelength of this light inside the lens.   **3**

  (*b*) The ray of red light is now replaced by a ray of blue light.

      The ray is incident on the lens at the same point as in part (*a*).

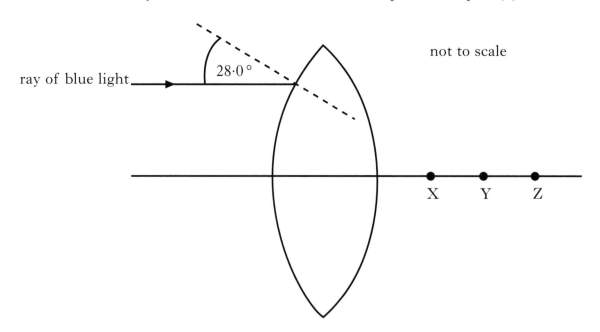

    Through which point, X, Y or Z, will this ray pass after leaving the lens?

    You must justify your answer.   **1**

                                                                (6)

Marks

28. The diagram shows a light sensor connected to a voltmeter.

A small lamp is placed in front of the sensor.

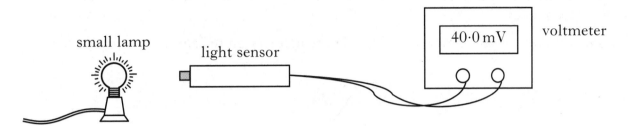

The reading on the voltmeter is 20 mV for each 1·0 mW of power incident on the sensor.

(a) The reading on the voltmeter is 40·0 mV.

The area of the light sensor is $8·0 \times 10^{-5} m^2$.

Calculate the irradiance of light on the sensor.   3

(b) The small lamp is replaced by a different source of light.

Using this new source, a student investigates how irradiance varies with distance.

The results are shown.

Distance/m	0·5	0·7	0·9
Irradiance/W m^{-2}	1·1	0·8	0·6

Can this new source be considered to be a point source of light?

Use **all** the data to justify your answer.   2

          (5)

[Turn over

Marks

29. To explain the photoelectric effect, light can be considered as consisting of tiny bundles of energy. These bundles of energy are called photons.

    (a) Sketch a graph to show the relationship between photon energy and frequency.   1

    (b) Photons of frequency 6.1×10^{14} Hz are incident on the surface of a metal.

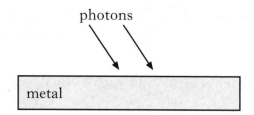

    This releases photoelectrons from the surface of the metal.

    The maximum kinetic energy of any of these photoelectrons is 6.0×10^{-20} J.

    Calculate the work function of the metal.   3

    (c) The irradiance due to these photons on the surface of the metal is now reduced.

    Explain why the maximum kinetic energy of each photoelectron is unchanged.   1

    (5)

Marks

30. (*a*) A technician is carrying out an experiment on the absorption of gamma radiation.

The radioactive source used has a long half-life and emits only gamma radiation. The activity of the source is 12 kBq.

(i) State what is meant by an *activity of 12 kBq*.   1

(ii) The table shows the half-value thicknesses of aluminium and lead for gamma radiation.

Material	Half-value thickness/mm
aluminium (Al)	60
lead (Pb)	15

The technician sets up the following apparatus.

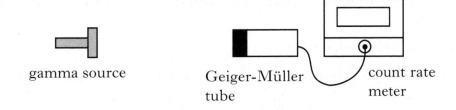

gamma source   Geiger-Müller   count rate
                      tube   meter

The count rate, when corrected for background radiation, is 800 counts per second.

Samples of aluminium and lead are now placed between the source and detector as shown.

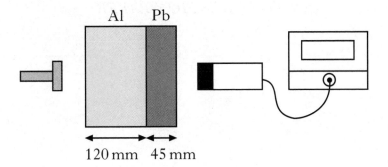

Al   Pb

120 mm   45 mm

Determine the new corrected count rate.   2

[Turn over for Question 30 (*b*) on *Page twenty-four*

Marks

30. (continued)

(b) X-ray scanners are used as part of airport security. A beam of X-rays scans the luggage as it passes through the scanner.

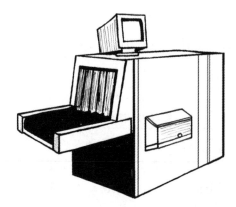

A baggage handler sometimes puts a hand inside the scanner to clear blockages.

The hand receives an average absorbed dose of $0{\cdot}030\,\mu$Gy each time this occurs.

The radiation weighting factor for X-rays is 1.

(i) State the average equivalent dose received by the hand on each occasion.   **1**

(ii) The occupational exposure limit for a hand is $60\,\mu$Sv per hour.

Calculate how many times the baggage handler would have to put a hand into the scanner in one hour to reach this limit.   **1**

(5)

[END OF QUESTION PAPER]

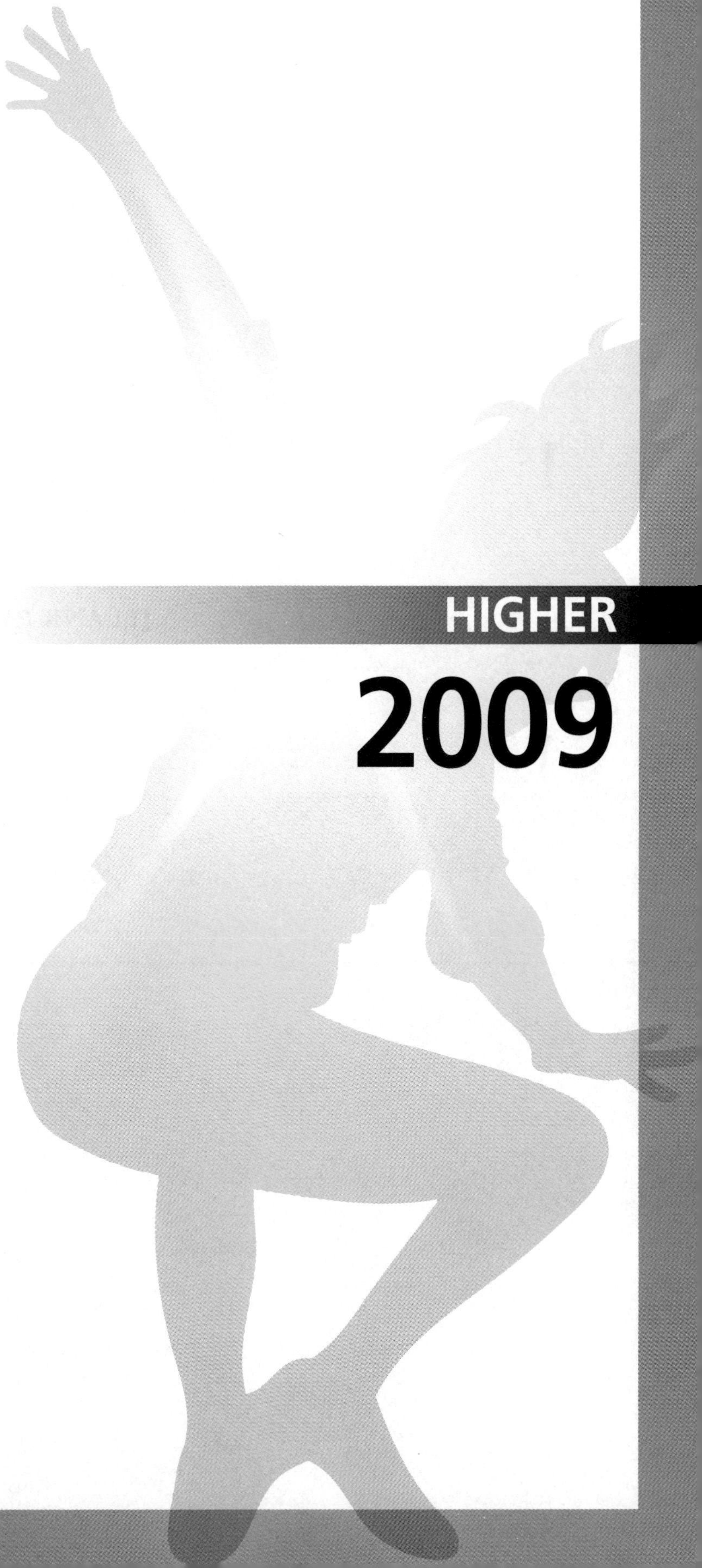

HIGHER

2009

[BLANK PAGE]

X069/301

NATIONAL QUALIFICATIONS 2009	TUESDAY, 26 MAY 1.00 PM – 3.30 PM	**PHYSICS HIGHER**

Read Carefully

Reference may be made to the Physics Data Booklet.

1 All questions should be attempted.

Section A (questions 1 to 20)

2 Check that the answer sheet is for Physics Higher (Section A).

3 For this section of the examination you must use an **HB pencil** and, where necessary, an eraser.

4 Check that the answer sheet you have been given has **your name**, **date of birth**, **SCN** (Scottish Candidate Number) and **Centre Name** printed on it.
Do not change any of these details.

5 If any of this information is wrong, tell the Invigilator immediately.

6 If this information is correct, **print** your name and seat number in the boxes provided.

7 There is **only one correct** answer to each question.

8 Any rough working should be done on the question paper or the rough working sheet, **not** on your answer sheet.

9 At the end of the exam, put the **answer sheet for Section A inside the front cover of your answer book**.

10 Instructions as to how to record your answers to questions 1–20 are given on page three.

Section B (questions 21 to 30)

11 Answer the questions numbered 21 to 30 in the answer book provided.

12 **All answers must be written clearly and legibly in ink.**

13 Fill in the details on the front of the answer book.

14 Enter the question number clearly in the margin of the answer book beside each of your answers to questions 21 to 30.

15 Care should be taken to give an appropriate number of significant figures in the final answers to calculations.

16 Where additional paper, eg square ruled paper, is used, write your name and SCN (Scottish Candidate Number) on it and place it inside the front cover of your answer booklet.

DATA SHEET
COMMON PHYSICAL QUANTITIES

Quantity	Symbol	Value	Quantity	Symbol	Value
Speed of light in vacuum	c	$3 \cdot 00 \times 10^{8}\,\text{m s}^{-1}$	Mass of electron	m_e	$9 \cdot 11 \times 10^{-31}\,\text{kg}$
Magnitude of the charge on an electron	e	$1 \cdot 60 \times 10^{-19}\,\text{C}$	Mass of neutron	m_n	$1 \cdot 675 \times 10^{-27}\,\text{kg}$
Gravitational acceleration on Earth	g	$9 \cdot 8\,\text{m s}^{-2}$	Mass of proton	m_p	$1 \cdot 673 \times 10^{-27}\,\text{kg}$
Planck's constant	h	$6 \cdot 63 \times 10^{-34}\,\text{J s}$			

REFRACTIVE INDICES
The refractive indices refer to sodium light of wavelength 589 nm and to substances at a temperature of 273 K.

Substance	Refractive index	Substance	Refractive index
Diamond	2·42	Water	1·33
Crown glass	1·50	Air	1·00

SPECTRAL LINES

Element	Wavelength/nm	Colour	Element	Wavelength/nm	Colour
Hydrogen	656	Red	Cadmium	644	Red
	486	Blue-green		509	Green
	434	Blue-violet		480	Blue
	410	Violet			
	397	Ultraviolet			
	389	Ultraviolet			
Sodium	589	Yellow			

			Lasers		
			Element	Wavelength/nm	Colour
			Carbon dioxide	9550 10590	Infrared
			Helium-neon	633	Red

PROPERTIES OF SELECTED MATERIALS

Substance	Density/ kg m^{-3}	Melting Point/ K	Boiling Point/ K
Aluminium	$2 \cdot 70 \times 10^{3}$	933	2623
Copper	$8 \cdot 96 \times 10^{3}$	1357	2853
Ice	$9 \cdot 20 \times 10^{2}$	273	. . . .
Sea Water	$1 \cdot 02 \times 10^{3}$	264	377
Water	$1 \cdot 00 \times 10^{3}$	273	373
Air	$1 \cdot 29$	. . . .	. . . .
Hydrogen	$9 \cdot 0 \ \times 10^{-2}$	14	20

The gas densities refer to a temperature of 273 K and a pressure of $1 \cdot 01 \times 10^{5}$ Pa.

SECTION A

For questions 1 to 20 in this section of the paper the answer to each question is either A, B, C, D or E. Decide what your answer is, then, using your pencil, put a horizontal line in the space provided—see the example below.

EXAMPLE

The energy unit measured by the electricity meter in your home is the

A kilowatt-hour

B ampere

C watt

D coulomb

E volt.

The correct answer is **A**—kilowatt-hour. The answer **A** has been clearly marked in **pencil** with a horizontal line (see below).

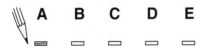

Changing an answer

If you decide to change your answer, carefully erase your first answer and, using your pencil, fill in the answer you want. The answer below has been changed to **E**.

[Turn over

SECTION A

Answer questions 1–20 on the answer sheet.

1. Which of the following contains one vector and one scalar quantity?

   A   power; speed

   B   force; kinetic energy

   C   momentum; velocity

   D   work; potential energy

   E   displacement; acceleration

2. The following velocity-time graph represents the vertical motion of a ball.

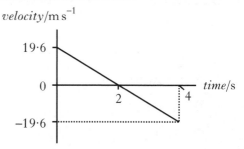

   Which of the following acceleration-time graphs represents the same motion?

   A

   B

   C

   D

   E

3. A box of weight 120 N is placed on a smooth horizontal surface.

A force of 20 N is applied to the box as shown.

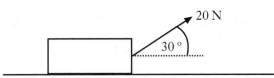

The box is pulled a distance of 50 m along the surface.

The work done in pulling the box is

A   500 J

B   866 J

C   1000 J

D   6000 J

E   6866 J.

4. A skydiver of total mass 85 kg is falling vertically.

At one point during the fall, the air resistance on the skydiver is 135 N.

The acceleration of the skydiver at this point is

A $0.6 \, \text{m s}^{-2}$

B $1.6 \, \text{m s}^{-2}$

C $6.2 \, \text{m s}^{-2}$

D $8.2 \, \text{m s}^{-2}$

E $13.8 \, \text{m s}^{-2}$.

5. A 2·0 kg trolley travels in a straight line towards a stationary 5·0 kg trolley as shown.

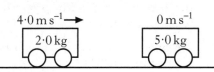

The trolleys collide. After the collision the trolleys move as shown below.

What is the speed v of the 5·0 kg trolley after the collision?

A $0.4 \, \text{m s}^{-1}$

B $1.2 \, \text{m s}^{-1}$

C $2.0 \, \text{m s}^{-1}$

D $2.2 \, \text{m s}^{-1}$

E $3.0 \, \text{m s}^{-1}$

6. The density of the gas in a container is initially $5.0 \, \text{kg m}^{-3}$.

Which of the following increases the density of the gas?

  I Raising the temperature of the gas without changing its mass or volume.

  II Increasing the mass of the gas without changing its volume or temperature.

  III Increasing the volume of the gas without changing its mass or temperature.

A   II only

B   III only

C   I and II only

D   II and III only

E   I, II and III

[Turn over

7. For a fixed mass of gas at constant volume

A   the pressure is directly proportional to temperature in °C

B   the pressure is inversely proportional to temperature in °C

C   the pressure is directly proportional to temperature in K

D   the pressure is inversely proportional to temperature in K

E   (pressure × temperature in K) is constant.

8. A potential difference, V, is applied between two metal plates. The plates are 0·15 m apart. A charge of +4·0 mC is released from rest at the positively charged plate as shown.

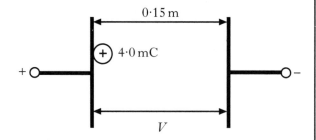

The kinetic energy of the charge just before it hits the negative plate is 8·0 J.

The potential difference between the plates is

A $3·2 \times 10^{-2}$ V

B   1·2 V

C   2·0 V

D $2·0 \times 10^{3}$ V

E $4·0 \times 10^{3}$ V.

9. A battery of e.m.f. 24 V and negligible internal resistance is connected as shown.

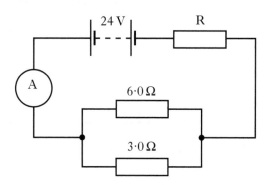

The reading on the ammeter is 2·0 A.

The resistance of R is

A   3·0 Ω

B   4·0 Ω

C   10 Ω

D   12 Ω

E   18 Ω.

10. The diagram shows a Wheatstone Bridge.

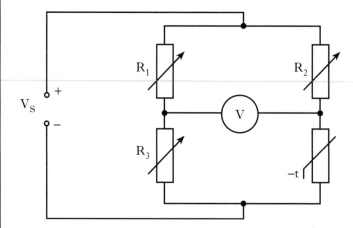

The bridge is initially balanced.

The thermistor is then heated and its resistance decreases. The bridge could be returned to balance by

A   decreasing R_1

B   decreasing R_2

C   increasing R_2

D   increasing R_3

E   increasing V_S.

11. A $25.0\,\mu F$ capacitor is charged until the potential difference across it is $500\,V$.

The charge stored in the capacitor is

A $5.00 \times 10^{-8}\,C$

B $2.00 \times 10^{-5}\,C$

C $1.25 \times 10^{-2}\,C$

D $1.25 \times 10^{4}\,C$

E $2.00 \times 10^{7}\,C$.

12. A student connects an a.c. supply to an a.c. ammeter and a component **X**.

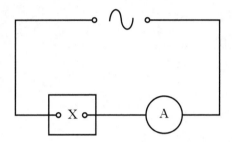

As the frequency of the a.c. supply is steadily increased, the ammeter reading also increases.

Component **X** is a

A   capacitor

B   diode

C   lamp

D   resistor

E   transistor.

13. An amplifier circuit is set up as shown.

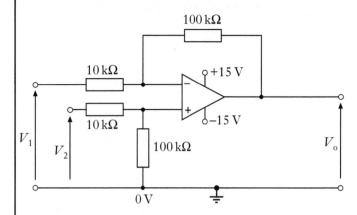

When $V_o = 0.60\,V$ and $V_1 = 2.70\,V$, what is the value of V_2?

A $2.10\,V$

B $2.64\,V$

C $2.76\,V$

D $3.30\,V$

E $8.70\,V$

[Turn over

14. A prism is used to produce a spectrum from a source of white light as shown.

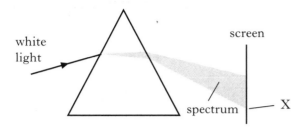

The colour observed at X is noted.

The prism is then replaced by a grating to produce spectra as shown.

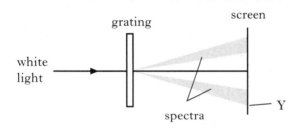

The colour observed at Y is noted.

Which row in the table gives the colour and wavelength of the light observed at X and the light observed at Y?

	Colour of light at X	Wavelength of light at X/nm	Colour of light at Y	Wavelength of light at Y/nm
A	Red	450	Red	450
B	Blue	450	Blue	450
C	Blue	650	Red	450
D	Blue	450	Red	650
E	Red	650	Blue	450

15. A ray of monochromatic light passes into a glass block as shown.

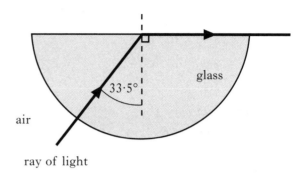

ray of light

The refractive index of the glass for this light is

A   0·03

B   0·55

C   0·87

D   1·20

E   1·81.

16. Which of the following statements about the characteristics of laser light is/are true?

I   It is monochromatic since all the photons have the same frequency.

II It is coherent because all the photons are in phase.

III Its irradiance is inversely proportional to the square of the distance from the source.

A   I only

B   I and II only

C   I and III only

D   II and III only

E   I, II and III

17. A student writes the following statements about p-type semiconductor material.

   I   Most charge carriers are positive.

   II   The p-type material has a positive charge.

   III   Impurity atoms in the material have 3 outer electrons.

   Which of these statements is/are true?

   A   I only

   B   II only

   C   I and II only

   D   I and III only

   E   I, II and III

18. A p-n junction diode is forward biased.

   Positive and negative charge carriers recombine in the junction region. This causes the emission of

   A   a hole

   B   an electron

   C   an electron-hole pair

   D   a proton

   E   a photon.

19. A sample of radioactive material has a mass of 20 g. There are 48 000 nuclear decays every minute in this sample.

   The activity of the sample is

   A   800 Bq

   B   2400 Bq

   C   48 000 Bq

   D   2 400 000 Bq

   E   2 880 000 Bq.

20. A sample of body tissue is irradiated by two different types of radiation, X and Y.

   The table gives the radiation weighting factor and absorbed dose for each radiation.

Type of radiation	Radiation weighting factor	Absorbed dose/μGy
X	10	5
Y	5	2

   The total equivalent dose received by the tissue is

   A $0.9\,\mu$Sv

   B $4.5\,\mu$Sv

   C $7.0\,\mu$Sv

   D $40.0\,\mu$Sv

   E $60.0\,\mu$Sv.

[Turn over

SECTION B

Write your answers to questions 21 to 30 in the answer book.

Marks

21. A basketball player throws a ball with an initial velocity of $6.5\,\text{m s}^{-1}$ at an angle of $50°$ to the horizontal. The ball is $2.3\,\text{m}$ above the ground when released.

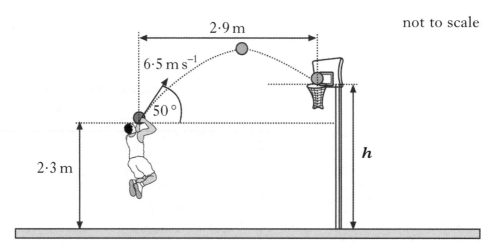

The ball travels a horizontal distance of $2.9\,\text{m}$ to reach the top of the basket.

The effects of air resistance can be ignored.

(a) Calculate:

    (i)   the horizontal component of the initial velocity of the ball;   **1**

    (ii)   the vertical component of the initial velocity of the ball.   **1**

(b) Show that the time taken for the ball to reach the basket is $0.69\,\text{s}$.   **1**

(c) Calculate the height **h** of the top of the basket.   **2**

(d) A student observing the player makes the following statement.

"The player should throw the ball with a higher speed at the same angle. The ball would then land in the basket as before but it would take a shorter time to travel the 2·9 metres."

Explain why the student's statement is incorrect.   **2**

    (7)

Marks

22. Golf clubs are tested to ensure they meet certain standards.

(*a*) In one test, a securely held clubhead is hit by a small steel pendulum. The time of contact between the clubhead and the pendulum is recorded.

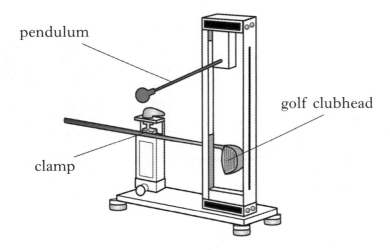

The experiment is repeated several times.

The results are shown.

248 μs   259 μs   251 μs   263 μs   254 μs

   (i)   Calculate:

       (A)   the mean contact time between the clubhead and the pendulum;   **1**

       (B)   the approximate absolute random uncertainty in this value.   **1**

   (ii)   In this test, the standard required is that the maximum value of the mean contact time must not be greater than 257 μs.

       Does the club meet this standard?

       You must justify your answer.   **1**

(*b*) In another test, a machine uses a club to hit a stationary golf ball.

The mass of the ball is $4 \cdot 5 \times 10^{-2}$ kg. The ball leaves the club with a speed of $50 \cdot 0 \, \text{m s}^{-1}$. The time of contact between the club and ball is 450 μs.

   (i)   Calculate the average force exerted on the ball by the club.   **2**

   (ii)   The test is repeated using a different club and an identical ball. The machine applies the same average force on the ball but with a longer contact time.

       What effect, if any, does this have on the speed of the ball as it leaves the club?

       Justify your answer.   **2**

(7)

Marks

23. A student is training to become a diver.

(*a*) The student carries out an experiment to investigate the relationship between the pressure and volume of a fixed mass of gas using the apparatus shown.

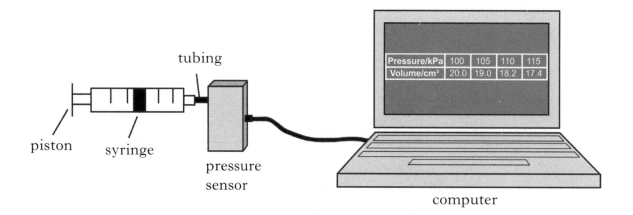

The pressure of the gas is recorded using a pressure sensor connected to a computer. The volume of the gas is also recorded. The student pushes the piston to alter the volume and a series of readings is taken.
The temperature of the gas is constant during the experiment.

The results are shown.

Pressure/kPa	100	105	110	115
Volume/cm³	20·0	19·0	18·2	17·4

  (i) Using **all** the data, establish the relationship between the pressure and volume of the gas. 2

  (ii) Use the kinetic model to explain the change in pressure as the volume of gas decreases. 2

(*b*) (i) The density of water in a loch is $1·02 \times 10^{3}\,\text{kg}\,\text{m}^{-3}$. Atmospheric pressure is $1·01 \times 10^{5}\,\text{Pa}$.

  Show that the **total** pressure at a depth of $12·0\,\text{m}$ in this loch is $2·21 \times 10^{5}\,\text{Pa}$. 2

  (ii) At the surface of the loch, the student breathes in a volume of $1·50 \times 10^{-3}\,\text{m}^{3}$ of air.

  Calculate the volume this air would occupy at a depth of $12·0\,\text{m}$. The mass and temperature of the air are constant. 2

(*c*) At a depth of $12·0\,\text{m}$, the diver fills her lungs with air from her breathing apparatus. She then swims to the surface.

  Explain why it would be dangerous for her to hold her breath while doing this. 2

(10)

Marks

24. A battery of e.m.f. 6·0 V and internal resistance, *r*, is connected to a variable resistor R as shown.

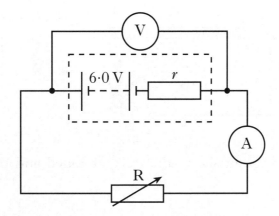

The graph shows how the current in the circuit changes as the resistance of R increases.

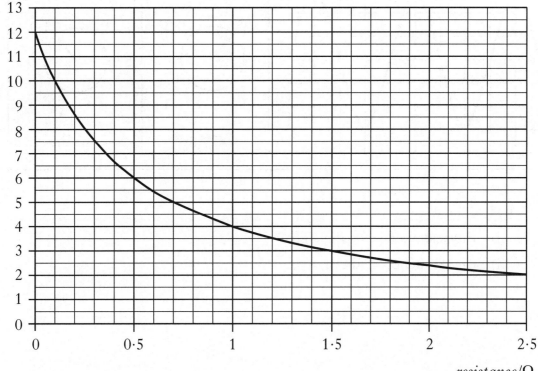

(a) Use information from the graph to calculate:

    (i) the lost volts in the circuit when the resistance of R is 1·5 Ω;   2

    (ii) the internal resistance, *r*, of the battery.   2

(b) The resistance of R is now increased.

    What effect, if any, does this have on the lost volts?

    You must justify your answer.   2

         (6)

Marks

25. (*a*) A microphone is connected to the input terminals of an oscilloscope.
A tuning fork is made to vibrate and held close to the microphone as shown.

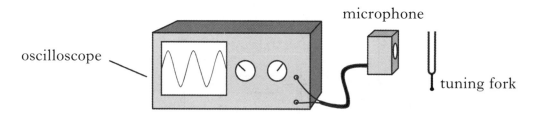

The following diagram shows the trace obtained and the settings on the oscilloscope.

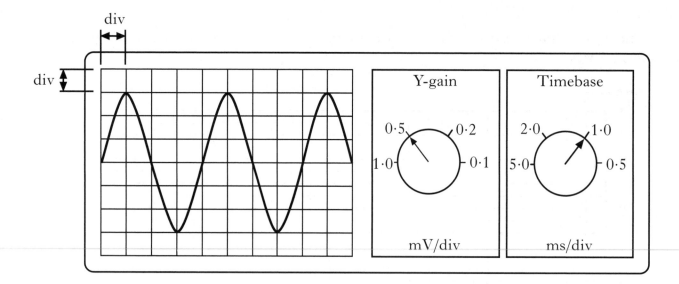

Calculate:

(i)   the peak voltage of the signal;   1

(ii) the frequency of the signal.   2

Marks

25. (continued)

(b) To amplify the signal from the microphone, it is connected to an op-amp circuit. The oscilloscope is now connected to the output of the amplifier as shown.

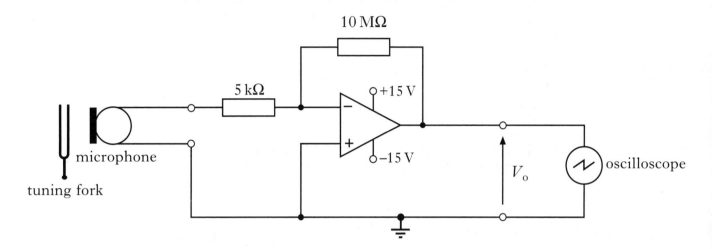

The settings of the oscilloscope are adjusted to show a trace of the amplified signal.

(i) In which mode is this op-amp being used? **1**

(ii) The peak voltage from the microphone is now 6·2 mV.

Calculate the **r.m.s.** value of the output voltage, V_o, of the op-amp. **3**

(iii) With the same input signal and settings on the oscilloscope, the supply voltage to the op-amp is now reduced from ± 15 V to ± 9 V.

What effect does this change have on the trace on the oscilloscope?

Justify your answer. **2**

(9)

[Turn over

Marks

26. A 12 volt battery of negligible internal resistance is connected in a circuit as shown.

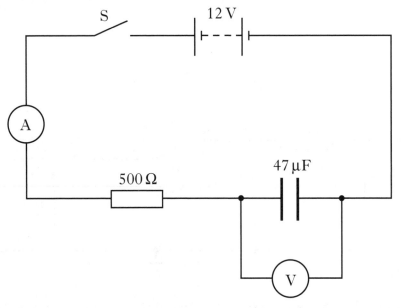

The capacitor is initially uncharged. Switch S is then closed and the capacitor starts to charge.

(a) Sketch a graph of the current against time from the instant switch S is closed. Numerical values are not required.   **1**

(b) At one instant during the charging of the capacitor the reading on the ammeter is 5·0 mA.

Calculate the reading on the voltmeter at this instant.   **3**

(c) Calculate the **maximum** energy stored in the capacitor in this circuit.   **2**

(d) The 500 Ω resistor is now replaced with a 2·0 kΩ resistor.

What effect, if any, does this have on the maximum energy stored in the capacitor?

Justify your answer.   **2**

(8)

Marks

27. A laser produces a narrow beam of monochromatic light.

(a) Red light from a laser passes through a grating as shown.

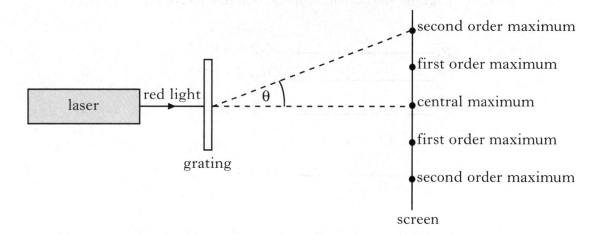

A series of maxima and minima is observed.

Explain in terms of waves how a **minimum** is produced.   1

(b) The laser is now replaced by a second laser, which emits blue light.

Explain why the observed maxima are now closer together.   1

(c) The wavelength of the blue light from the second laser is $4.73 \times 10^{-7}\,\text{m}$. The spacing between the lines on the grating is $2.00 \times 10^{-6}\,\text{m}$.

Calculate the angle between the central maximum and the second order maximum.   2

(4)

[Turn over

Marks

28. (a) Electrons which orbit the nucleus of an atom can be considered as occupying discrete energy levels.

The following diagram shows some of the energy levels for a particular atom.

E_3 ———————————— $-5{\cdot}2 \times 10^{-19}\,\text{J}$

E_2 ———————————— $-9{\cdot}0 \times 10^{-19}\,\text{J}$

E_1 ———————————— $-16{\cdot}2 \times 10^{-19}\,\text{J}$

E_0 ———————————— $-24{\cdot}6 \times 10^{-19}\,\text{J}$

(i) Radiation is produced when electrons make transitions from a higher to a lower energy level.

Which transition, between these energy levels, produces radiation with the shortest wavelength?

Justify your answer.   2

(ii) An electron is excited from energy level E_2 to E_3 by absorbing light energy.

What frequency of light is used to excite this electron?   2

(b) Another source of light has a frequency of $4{\cdot}6 \times 10^{14}\,\text{Hz}$ in air.

A ray of this light is directed into a block of transparent material as shown.

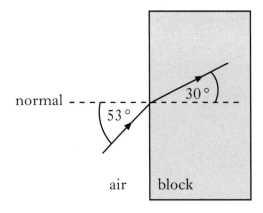

Calculate the wavelength of the light in the block.   3

(7)

Marks

29. Ultraviolet radiation from a lamp is incident on the surface of a metal.

This causes the release of electrons from the surface of the metal.

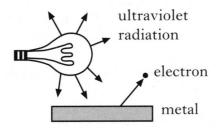

The energy of each photon of ultraviolet light is $5 \cdot 23 \times 10^{-19}$ J.

The work function of the metal is $2 \cdot 56 \times 10^{-19}$ J.

(a) Calculate:

    (i)   the maximum kinetic energy of an electron released from this metal by this radiation;   **1**

    (ii) the maximum speed of an emitted electron.   **2**

(b) The source of ultraviolet radiation is now moved further away from the surface of the metal.

State the effect, if any, this has on the maximum speed of an emitted electron.

Justify your answer.   **2**

(5)

[Turn over

Marks

30. (*a*) Some power stations use nuclear fission reactions to provide energy for generating electricity. The following statement represents a fission reaction.

$$^{235}_{92}U + ^{1}_{0}n \rightarrow ^{139}_{57}La + ^{r}_{42}Mo + 2^{1}_{0}n + s^{0}_{-1}e$$

(i) Determine the numbers represented by the letters *r* and *s* in the above statement.   **1**

(ii) Explain why a nuclear fission reaction releases energy.   **1**

(iii) The masses of the particles involved in the reaction are shown in the table.

Particle	Mass/kg
$^{235}_{92}U$	$390 \cdot 173 \times 10^{-27}$
$^{139}_{57}La$	$230 \cdot 584 \times 10^{-27}$
$^{r}_{42}Mo$	$157 \cdot 544 \times 10^{-27}$
$^{1}_{0}n$	$1 \cdot 675 \times 10^{-27}$
$^{0}_{-1}e$	negligible

Calculate the energy released in this reaction.   **3**

Marks

30. (continued)

(b) One method of reducing the radiation received by a person is by using lead shielding.

In an investigation of the absorption of gamma radiation by lead, the following graph of corrected count rate against thickness of lead is obtained.

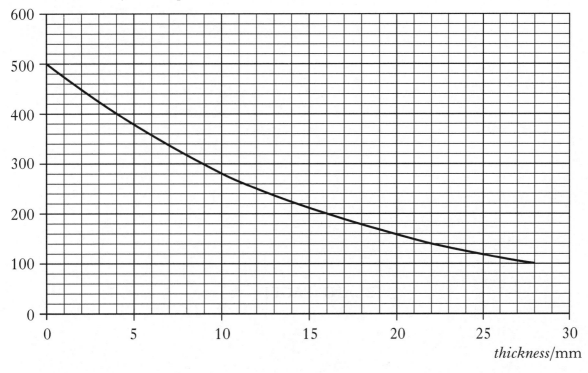

corrected count rate/counts per minute

thickness/mm

   (i) Determine the half-value thickness of lead for this radiation. **1**

  (ii) With no shielding, the equivalent dose rate a short distance from this source is $200\,\mu Sv\,h^{-1}$.

When the source is stored in a lead container, the equivalent dose rate at the same distance falls to $50\,\mu Sv\,h^{-1}$.

Calculate the thickness of the lead container. **1**

(7)

[*END OF QUESTION PAPER*]

[BLANK PAGE]

HIGHER

2010

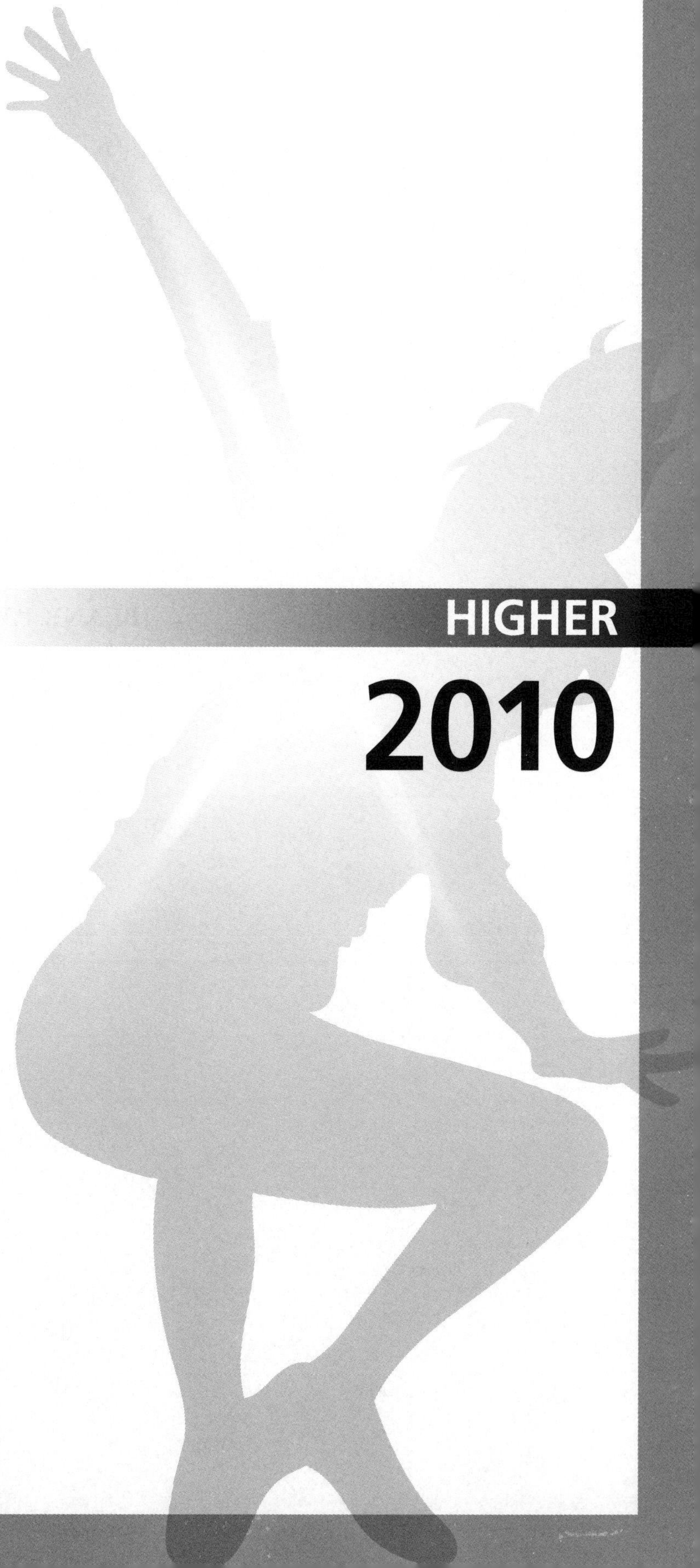

[BLANK PAGE]

X069/301

NATIONAL QUALIFICATIONS 2010	FRIDAY, 28 MAY 1.00 PM – 3.30 PM	PHYSICS HIGHER

Read Carefully

Reference may be made to the Physics Data Booklet.

1 All questions should be attempted.

Section A (questions 1 to 20)

2 Check that the answer sheet is for Physics Higher (Section A).

3 For this section of the examination you must use an **HB pencil** and, where necessary, an eraser.

4 Check that the answer sheet you have been given has **your name**, **date of birth**, **SCN** (Scottish Candidate Number) and **Centre Name** printed on it.

   Do not change any of these details.

5 If any of this information is wrong, tell the Invigilator immediately.

6 If this information is correct, **print** your name and seat number in the boxes provided.

7 There is **only one correct** answer to each question.

8 Any rough working should be done on the question paper or the rough working sheet, **not** on your answer sheet.

9 At the end of the exam, put the **answer sheet for Section A inside the front cover of your answer book**.

10 Instructions as to how to record your answers to questions 1–20 are given on page three.

Section B (questions 21 to 30)

11 Answer the questions numbered 21 to 30 in the answer book provided.

12 **All answers must be written clearly and legibly in ink.**

13 Fill in the details on the front of the answer book.

14 Enter the question number clearly in the margin of the answer book beside each of your answers to questions 21 to 30.

15 Care should be taken to give an appropriate number of significant figures in the final answers to calculations.

16 Where additional paper, eg square ruled paper, is used, write your name and SCN (Scottish Candidate Number) on it and place it inside the front cover of your answer booklet.

DATA SHEET
COMMON PHYSICAL QUANTITIES

Quantity	Symbol	Value	Quantity	Symbol	Value
Speed of light in vacuum	c	3.00×10^8 m s^{-1}	Mass of electron	m_e	9.11×10^{-31} kg
Magnitude of the charge on an electron	e	1.60×10^{-19} C	Mass of neutron	m_n	1.675×10^{-27} kg
Gravitational acceleration on Earth	g	9.8 m s^{-2}	Mass of proton	m_p	1.673×10^{-27} kg
Planck's constant	h	6.63×10^{-34} J s			

REFRACTIVE INDICES

The refractive indices refer to sodium light of wavelength 589 nm and to substances at a temperature of 273 K.

Substance	Refractive index	Substance	Refractive index
Diamond	2.42	Water	1.33
Crown glass	1.50	Air	1.00

SPECTRAL LINES

Element	Wavelength/nm	Colour	Element	Wavelength/nm	Colour
Hydrogen	656	Red	Cadmium	644	Red
	486	Blue-green		509	Green
	434	Blue-violet		480	Blue
	410	Violet	Lasers		
	397	Ultraviolet			
	389	Ultraviolet	Element	Wavelength/nm	Colour
Sodium	589	Yellow	Carbon dioxide	9550 } 10590 }	Infrared
			Helium-neon	633	Red

PROPERTIES OF SELECTED MATERIALS

Substance	Density/ kg m^{-3}	Melting Point/ K	Boiling Point/ K
Aluminium	2.70×10^3	933	2623
Copper	8.96×10^3	1357	2853
Ice	9.20×10^2	273	. . . .
Sea Water	1.02×10^3	264	377
Water	1.00×10^3	273	373
Air	1.29	. . . .	. . . .
Hydrogen	9.0×10^{-2}	14	20

The gas densities refer to a temperature of 273 K and a pressure of 1.01×10^5 Pa.

SECTION A

For questions 1 to 20 in this section of the paper the answer to each question is either A, B, C, D or E. Decide what your answer is, then, using your pencil, put a horizontal line in the space provided—see the example below.

EXAMPLE

The energy unit measured by the electricity meter in your home is the

    A kilowatt-hour

    B ampere

    C watt

    D coulomb

    E volt.

The correct answer is **A**—kilowatt-hour. The answer **A** has been clearly marked in **pencil** with a horizontal line (see below).

Changing an answer

If you decide to change your answer, carefully erase your first answer and, using your pencil, fill in the answer you want. The answer below has been changed to **E**.

[Turn over

SECTION A

Answer questions 1–20 on the answer sheet.

1. Acceleration is the change in

   A   distance per unit time

   B   displacement per unit time

   C   velocity per unit distance

   D   speed per unit time

   E   velocity per unit time.

2. The graph shows how the acceleration, a, of an object varies with time, t.

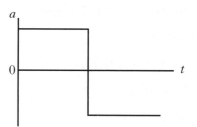

   Which graph shows how the velocity, v, of the object varies with time, t?

   A
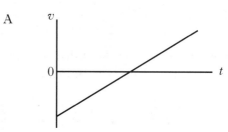

   B
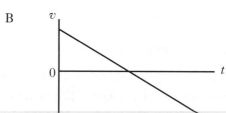

   C
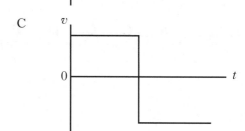

   D
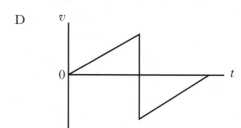

   E
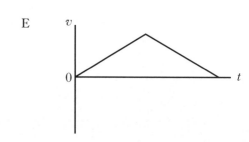

3. A car of mass 1000 kg is travelling at a speed of 40 m s⁻¹ along a race track. The brakes are applied and the speed of the car decreases to 10 m s⁻¹.

   How much kinetic energy is lost by the car?

   A   15 kJ

   B   50 kJ

   C   450 kJ

   D   750 kJ

   E   800 kJ

4. A substance can exist as a solid, a liquid or a gas.

   Which row in the table shows the approximate relative magnitudes of the densities of the substance in these states?

	Density of solid	Density of liquid	Density of gas
A	1000	1000	1
B	10	10	1000
C	1	1	1000
D	1000	10	1
E	1	1	10

5. A fish is swimming at a depth of 10·4 m.

   The density of the water is $1·03 \times 10^3 \, kg\, m^{-3}$.

   The pressure at this depth caused by the water is

   A $0·99 \times 10^2 \, Pa$

   B $1·04 \times 10^4 \, Pa$

   C $1·07 \times 10^4 \, Pa$

   D $1·05 \times 10^5 \, Pa$

   E $1·07 \times 10^5 \, Pa$.

6. Ice at a temperature of −10 °C is heated until it becomes water at 80 °C.

   The temperature change in kelvin is

   A   70 K

   B   90 K

   C   343 K

   D   363 K

   E   636 K.

7. The potential difference between two points is

   A   the work done in moving one electron between the two points

   B   the voltage between the two points when there is a current of one ampere

   C   the work done in moving one coulomb of charge between the two points

   D   the kinetic energy gained by an electron as it moves between the two points

   E   the work done in moving any charge between the two points.

8. The product, X, of a nuclear reaction passes through an electric field as shown.

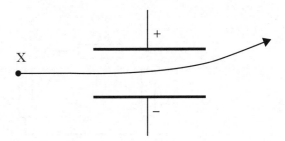

   Product X is

   A   an alpha particle

   B   a beta particle

   C   gamma radiation

   D   a fast neutron

   E   a slow neutron.

[Turn over

9. Which of the following combinations of resistors has the greatest resistance between X and Y?

A

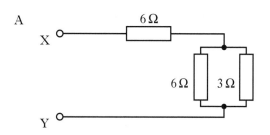

B

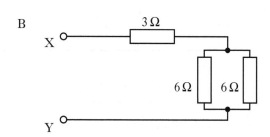

C

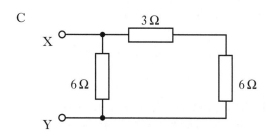

D

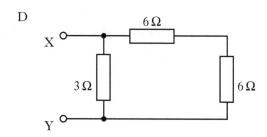

E

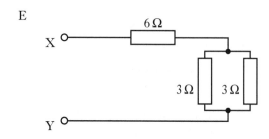

10. In the following Wheatstone bridge circuit, the reading on the voltmeter is zero when the resistance of R is set at $1\,k\Omega$.

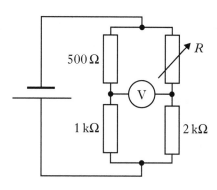

Which of the following is the graph of the voltmeter reading V against the resistance R?

A

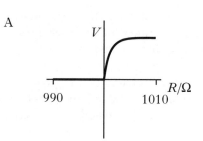

B

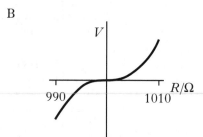

C

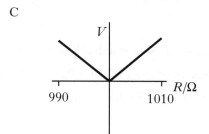

D

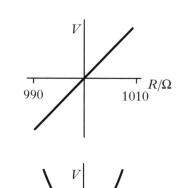

E

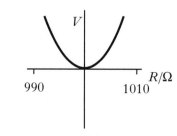

11. A student makes the following statements about capacitors.

    I Capacitors block a.c. signals.

    II Capacitors store energy.

    III Capacitors store charge.

    Which of these statements is/are true?

    A I only

    B I and II only

    C I and III only

    D II and III only

    E I, II and III

12. A circuit is set up as shown.

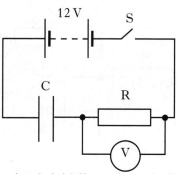

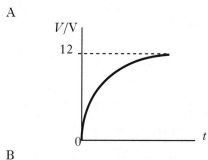

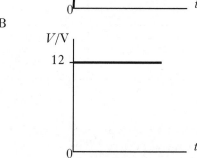

The capacitor is initially uncharged. Switch S is now closed. Which graph shows how the potential difference, V, across R, varies with time, t?

A

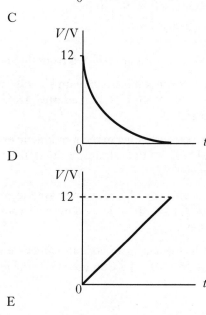

B

C

D

E

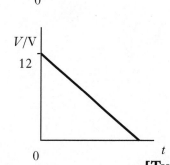

13. An op-amp is connected in a circuit as shown.

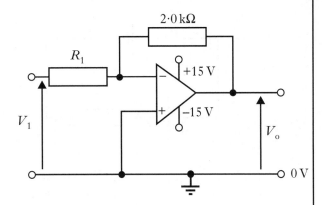

The input voltage V_1 is $0.50\,V$.

Which row in the table shows possible values for R_1 and V_o?

	$R_1/k\Omega$	V_o/V
A	1.0	1.0
B	4.0	1.0
C	1.0	-0.25
D	4.0	-1.0
E	1.0	-1.0

14. Photons of energy $7.0 \times 10^{-19}\,J$ are incident on a clean metal surface. The work function of the metal is $9.0 \times 10^{-19}\,J$.

Which of the following is correct?

A   No electrons are emitted from the metal.

B   Electrons with a maximum kinetic energy of $2.0 \times 10^{-19}\,J$ are emitted from the metal.

C   Electrons with a maximum kinetic energy of $7.0 \times 10^{-19}\,J$ are emitted from the metal.

D   Electrons with a maximum kinetic energy of $9.0 \times 10^{-19}\,J$ are emitted from the metal.

E   Electrons with a maximum kinetic energy of $16 \times 10^{-19}\,J$ are emitted from the metal.

15. The diagram represents some of the energy levels for an atom of a gas.

E_3 ——————— $-5.2 \times 10^{-19}\,J$
E_2 ——————— $-8.3 \times 10^{-19}\,J$

E_1 ——————— $-12.5 \times 10^{-19}\,J$

E_0 ——————— $-17.9 \times 10^{-19}\,J$

White light passes through the gas and absorption lines are observed in the spectrum.

Which electron transition produces the absorption line corresponding to the lowest frequency?

A E_3 to E_2

B E_2 to E_3

C E_1 to E_0

D E_0 to E_1

E E_0 to E_3

16. An LED is connected as shown.

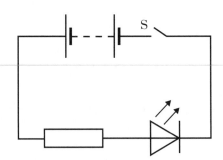

When switch S is closed

A   the p-n junction is reverse biased and free charge carriers are produced which may recombine to give quanta of radiation

B   the p-n junction is forward biased and positive and negative charge carriers are produced by the action of light

C   the p-n junction is reverse biased and positive and negative charge carriers are produced by the action of light

D   the p-n junction is forward biased and positive and negative charge carriers may recombine to give quanta of radiation

E   the p-n junction is reverse biased and positive and negative charge carriers may recombine to give quanta of radiation.

17. The diagram represents the structure of an n-channel enhancement MOSFET.

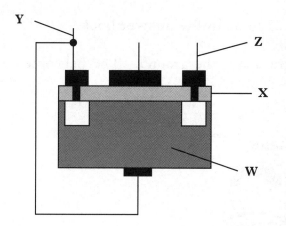

Which row in the table gives the names for the parts labelled **W**, **X**, **Y** and **Z**?

	W	**X**	**Y**	**Z**
A	substrate	implant	source	drain
B	implant	substrate	source	drain
C	substrate	oxide layer	drain	source
D	implant	substrate	gate	source
E	substrate	oxide layer	source	drain

18. The following statement describes a fusion reaction.

$$^2_1H + ^2_1H \longrightarrow ^3_2He + ^1_0n + energy$$

The total mass of the particles before the reaction is $6·684 \times 10^{-27}$ kg.

The total mass of the particles after the reaction is $6·680 \times 10^{-27}$ kg.

The energy released in this reaction is

A $6·012 \times 10^{-10}$ J

B $6·016 \times 10^{-10}$ J

C $1·800 \times 10^{-13}$ J

D $3·600 \times 10^{-13}$ J

E $1·200 \times 10^{-21}$ J.

19. A sample of tissue receives an equivalent dose of 40 mSv from a beam of neutrons.

The neutrons have a radiation weighting factor of 10.

The energy absorbed by the tissue is 100 μJ.

The mass of the tissue is

A $2·5 \times 10^{-4}$ kg

B $2·5 \times 10^{-2}$ kg

C $4·0$ kg

D 40 kg

E $4·0 \times 10^3$ kg.

20. A sample of tissue is placed near a source of gamma radiation. The equivalent dose rate for the tissue is 80 μSv h^{-1}.

The equivalent dose rate is now reduced to 10 μSv h^{-1} by placing lead shielding between the source and the tissue.

The half value thickness of lead is 8·0 mm for this source.

The thickness of the lead shielding is

A 1·0 mm

B 8·0 mm

C 24 mm

D 64 mm

E 80 mm.

[Turn over

Marks

SECTION B

Write your answers to questions 21 to 30 in the answer book.

21. A helicopter is flying at a constant height above the ground. The helicopter is carrying a crate suspended from a cable as shown.

(a) The helicopter flies 20 km on a bearing of 180 (due South). It then turns on to a bearing of 140 (50° South of East) and travels a further 30 km.

The helicopter takes 15 minutes to travel the 50 km.

   (i) By scale drawing (or otherwise) find the resultant displacement of the helicopter.   **2**

   (ii) Calculate the average velocity of the helicopter during the 15 minutes.   **2**

(b) The helicopter reaches its destination and hovers above a drop zone.

   (i) The total mass of the helicopter and crate is $1{\cdot}21 \times 10^4$ kg.

   Show that the helicopter produces a lift force of 119 kN.   **1**

   (ii) The helicopter now drops the crate which has a mass of $2{\cdot}30 \times 10^3$ kg. The lift force remains constant.

   Describe the vertical motion of the helicopter immediately after the crate is dropped.

   Justify your answer in terms of the forces acting on the helicopter.   **2**

    (7)

Marks

22. The apparatus shown is set up to investigate collisions between two vehicles on a track.

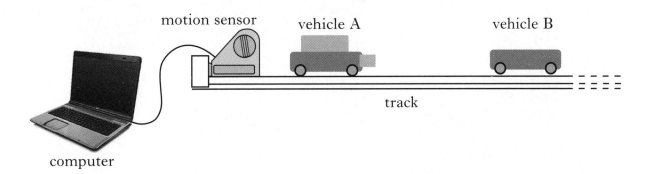

The mass of vehicle A is 0·22 kg and the mass of vehicle B is 0·16 kg.

The effects of friction are negligible.

(a) During one experiment the vehicles collide and stick together. The computer connected to the motion sensor displays the velocity-time graph for vehicle A.

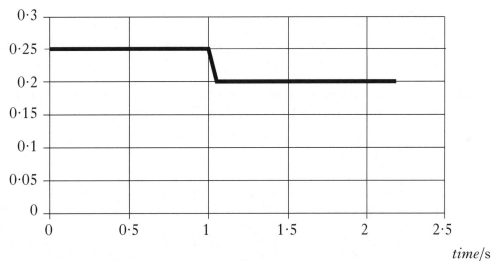

    (i) State the law of conservation of momentum.   **1**

    (ii) Calculate the velocity of vehicle B before the collision.   **2**

(b) The same apparatus is used to carry out a second experiment.

In this experiment, vehicle B is stationary before the collision.

Vehicle A has the same velocity before the collision as in the first experiment.

After the collision, the two vehicles stick together.

Is their combined velocity less than, equal to, or greater than that in the first collision?

Justify your answer.   **2**

    (5)

Marks

23. (*a*) A gymnast of mass 40 kg is practising on a trampoline.

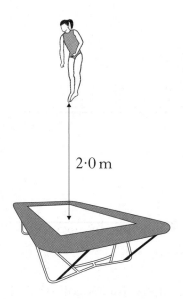

2·0 m

   (i) At maximum height the gymnast's feet are 2·0 m above the trampoline. Show that the speed of the gymnast, as she lands on the trampoline, is 6·3 m s^{-1}.   **1**

 (ii) The gymnast rebounds with a speed of 5·7 m s^{-1}. Calculate the change in momentum of the gymnast.   **2**

(iii) The gymnast was in contact with the trampoline for 0·50 s. Calculate the average force exerted by the trampoline on the gymnast.   **2**

Marks

23. (continued)

(*b*) Another gymnast is practising on a piece of equipment called the rings. The gymnast grips two wooden rings suspended above the gym floor by strong, vertical ropes as shown in Figure 1.

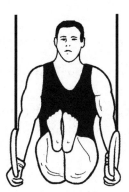

Figure 1

He now stretches out his arms until each rope makes an angle of 10° with the vertical as shown in Figure 2.

Figure 2

Explain why the tension in each rope increases as the gymnast stretches out his arms.

2

(7)

[Turn over

Marks

24. An experiment is carried out to measure the time taken for a steel ball to fall vertically through a fixed distance using an electronic timer.

   (*a*) The experiment is repeated and the following values for time recorded.

   0·49 s,   0·53 s,   0·50 s,   0·50 s,   0·55 s,   0·51 s.

   Calculate:

   (i)   the mean value of the time;   1

   (ii) the approximate random uncertainty in the mean value of the time.   1

   (*b*) Part of the circuit in the electronic timer consists of a 1·6 mF capacitor and an 18 kΩ resistor connected to a switch and a 4·5 V supply.

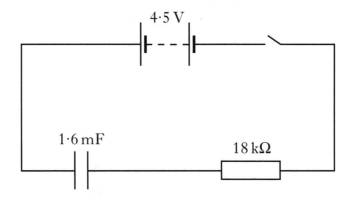

   (i)   Calculate the charge on the capacitor when it is fully charged.   2

   (ii) Sketch the graph of the current in the resistor against time as the capacitor charges.

   Numerical values are required on the current axis.   2

                                                                                               (6)

Marks

25. The headlights on a truck are switched on automatically when a light sensor detects the light level falling below a certain value.

The light sensor consists of an LDR connected in a Wheatstone bridge as shown.

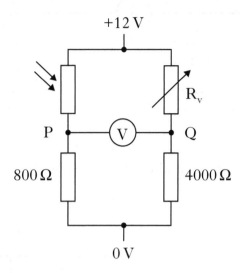

(a) The variable resistor, R_v, is set at $6000\,\Omega$.

  (i) Calculate the resistance of the LDR when the bridge is balanced.   2

  (ii) As the light level decreases, the resistance of the LDR increases. Calculate the reading on the voltmeter when the resistance of the LDR is $1600\,\Omega$.   2

(b) The Wheatstone bridge is connected to an op-amp as shown. The output of the op-amp controls the headlights circuit.

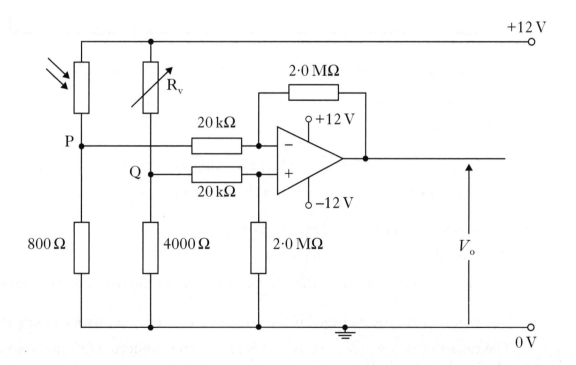

The resistance of R_v is adjusted so that the potential at Q is $3\cdot2\,V$. At a particular light level, the potential at P is $3\cdot0\,V$. Determine the output voltage, V_o, of the op-amp.   3

(7)

Marks

26. A signal generator is connected to a lamp, a resistor and an ammeter in series. An oscilloscope is connected across the output terminals of the signal generator.

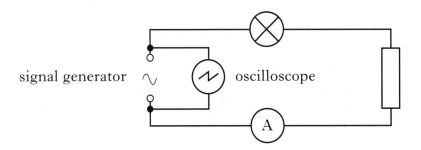

The oscilloscope control settings and the trace displayed on its screen are shown.

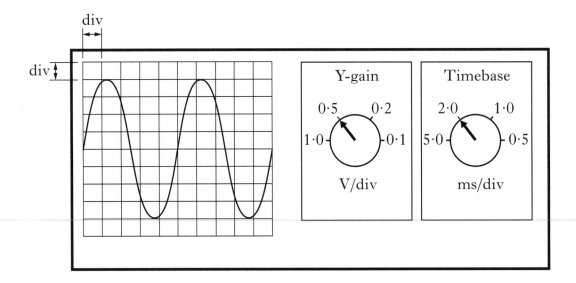

(a) For this signal calculate:

    (i) the peak voltage;   1

    (ii) the frequency.   2

(b) The frequency is now doubled. The peak voltage of the signal is kept constant.

    State what happens to the reading on the ammeter.   1

(c) The resistor is now replaced by a capacitor.

    The procedure in part (b) is repeated.

    State what happens to the reading on the ammeter as the frequency is doubled.   1

(d) The capacitor will be damaged if the potential difference across it exceeds 16 V.

    The capacitor is now removed from this circuit and connected to a different a.c. supply of output $15\,V_{r.m.s.}$.

    Explain whether or not the capacitor is damaged.   2

                                             (7)

Marks

27. A student is carrying out an experiment to investigate the interference of sound waves. She sets up the following apparatus.

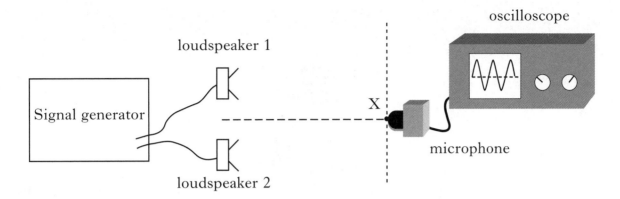

The microphone is initially placed at point X which is the same distance from each loudspeaker. A maximum is detected at X.

(a) The microphone is now moved to the first minimum at Y as shown.

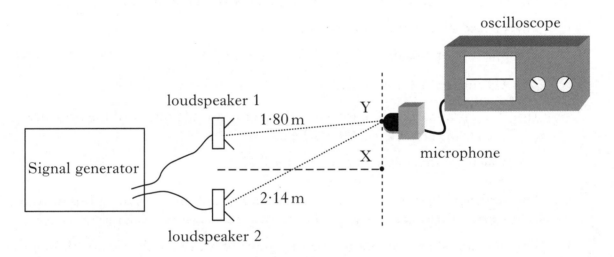

    Calculate the wavelength of the sound waves.   2

(b) Loudspeaker 1 is now disconnected.

    What happens to the amplitude of the sound detected by the microphone at Y?

    Explain your answer.   2

                                           (4)

[Turn over

Marks

28. A garden spray consists of a tank, a pump and a spray nozzle.

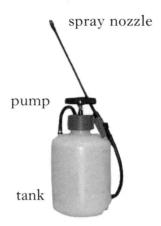

spray nozzle

pump

tank

The tank is partially filled with water.

The pump is then used to increase the pressure of the air above the water.

(a) The volume of the compressed air in the tank is $1\cdot60 \times 10^{-3}\,\text{m}^3$.

The surface area of the water is $3\cdot00 \times 10^{-2}\,\text{m}^2$.

The pressure of the air in the tank is $4\cdot60 \times 10^5\,\text{Pa}$.

(i) Calculate the force on the surface of the water.   2

(ii) The spray nozzle is operated and water is pushed out until the pressure of the air in the tank is $1\cdot00 \times 10^5\,\text{Pa}$.

Calculate the volume of water expelled.   3

(b) The gardener observes a spectrum when sunlight illuminates the drops of water in the spray. This is because each drop of water is acting as a prism.

The diagram shows the path taken by light of wavelength 650 nm through a drop of water.

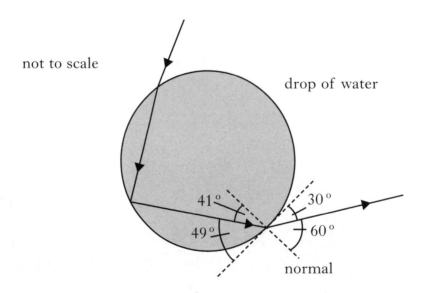

not to scale

drop of water

41°

49°

30°

60°

normal

(i) What happens to the frequency of this light when it enters the drop of water?   1

Marks

28. (*b*) (continued)

    (ii) Using information from the diagram, calculate the refractive index of the water for this wavelength of light.   **2**

    (iii) Calculate the critical angle for this wavelength of light in the water.   **2**

    (iv) Light of shorter wavelength also passes through the drop of water.

        Will the critical angle for this light be less than, equal to, or greater than that for light of wavelength 650 nm?

        Justify your answer.   **2**

        (12)

[Turn over

Marks

29. A laser produces a beam of light with a frequency of $4 \cdot 74 \times 10^{14}$ Hz.

(a) The laser has a power of $0 \cdot 10$ mW. Explain why light from this laser can cause eye damage.   1

(b) Calculate the energy of each photon in the laser beam.   2

(c) Inside the laser, photons stimulate the emission of more photons.

State **two** ways in which the stimulated photons are identical to the photons producing them.   1

(d) This laser beam is now incident on a grating as shown below.

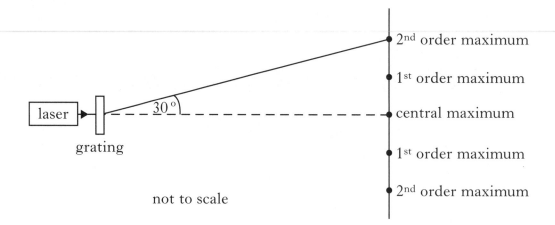

The second order maximum is detected at an angle of $30°$ from the central maximum.

Calculate the separation of the slits on the grating.   3

                                                                                                                                            (7)

Marks

30. A smoke alarm contains a very small sample of the radioactive isotope Americium-241, represented by the symbol

$$^{241}_{\ 95}\text{Am}$$

(a) How many neutrons are there in a nucleus of this isotope?   **1**

(b) This isotope decays by emitting alpha particles as shown in the following statement.

$$^{241}_{\ 95}\text{Am} \longrightarrow {}^{s}_{r}T + \alpha$$

    (i) Determine the numbers represented by the letters *r* and *s*.   **1**

    (ii) Use the data booklet to identify the element *T*.   **1**

(c) The activity of the radioactive sample is 30 kBq. How many decays take place in one minute?   **2**

(d) The alarm circuit in the smoke detector contains a battery of e.m.f. 9·0 V and internal resistance 2·0 Ω.

This circuit is shown.

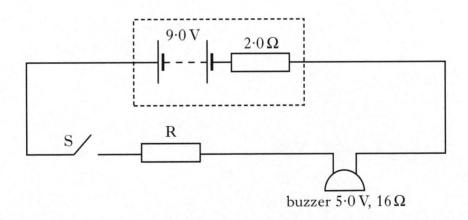

When smoke is detected, switch S closes and the buzzer operates. The buzzer has a resistance of 16 Ω and an operating voltage of 5·0 V.

Calculate the value of resistor R required in this circuit.   **3**

          (8)

[END OF QUESTION PAPER]

[BLANK PAGE]

2011

[BLANK PAGE]

X069/301

NATIONAL QUALIFICATIONS 2011	MONDAY, 23 MAY 1.00 PM – 3.30 PM	PHYSICS HIGHER

Read Carefully

Reference may be made to the Physics Data Booklet.

1   All questions should be attempted.

Section A (questions 1 to 20)

2   Check that the answer sheet is for Physics Higher (Section A).

3   For this section of the examination you must use an **HB pencil** and, where necessary, an eraser.

4   Check that the answer sheet you have been given has **your name**, **date of birth**, **SCN** (Scottish Candidate Number) and **Centre Name** printed on it.

Do not change any of these details.

5   If any of this information is wrong, tell the Invigilator immediately.

6   If this information is correct, **print** your name and seat number in the boxes provided.

7   There is **only one correct** answer to each question.

8   Any rough working should be done on the question paper or the rough working sheet, **not** on your answer sheet.

9   At the end of the exam, put the **answer sheet for Section A inside the front cover of your answer book**.

10   Instructions as to how to record your answers to questions 1–20 are given on page three.

Section B (questions 21 to 30)

11   Answer the questions numbered 21 to 30 in the answer book provided.

12   **All answers must be written clearly and legibly in ink**.

13   Fill in the details on the front of the answer book.

14   Enter the question number clearly in the margin of the answer book beside each of your answers to questions 21 to 30.

15   Care should be taken to give an appropriate number of significant figures in the final answers to calculations.

16   Where additional paper, eg square ruled paper, is used, write your name and SCN (Scottish Candidate Number) on it and place it inside the front cover of your answer booklet.

DATA SHEET
COMMON PHYSICAL QUANTITIES

Quantity	Symbol	Value	Quantity	Symbol	Value
Speed of light in vacuum	c	$3.00 \times 10^{8}\,\text{m s}^{-1}$	Mass of electron	m_e	$9.11 \times 10^{-31}\,\text{kg}$
Magnitude of the charge on an electron	e	$1.60 \times 10^{-19}\,\text{C}$	Mass of neutron	m_n	$1.675 \times 10^{-27}\,\text{kg}$
Gravitational acceleration on Earth	g	$9.8\,\text{m s}^{-2}$	Mass of proton	m_p	$1.673 \times 10^{-27}\,\text{kg}$
Planck's constant	h	$6.63 \times 10^{-34}\,\text{J s}$			

REFRACTIVE INDICES

The refractive indices refer to sodium light of wavelength 589 nm and to substances at a temperature of 273 K.

Substance	Refractive index	Substance	Refractive index
Diamond	2·42	Water	1·33
Crown glass	1·50	Air	1·00

SPECTRAL LINES

Element	Wavelength/nm	Colour	Element	Wavelength/nm	Colour
Hydrogen	656	Red	Cadmium	644	Red
	486	Blue-green		509	Green
	434	Blue-violet		480	Blue
	410	Violet			
	397	Ultraviolet		*Lasers*	
	389	Ultraviolet	Element	Wavelength/nm	Colour
			Carbon dioxide	9550 } 10590	Infrared
Sodium	589	Yellow	Helium-neon	633	Red

PROPERTIES OF SELECTED MATERIALS

Substance	Density/kg m^{-3}	Melting Point/K	Boiling Point/K
Aluminium	2.70×10^{3}	933	2623
Copper	8.96×10^{3}	1357	2853
Ice	9.20×10^{2}	273	. . . .
Sea Water	1.02×10^{3}	264	377
Water	1.00×10^{3}	273	373
Air	1·29	. . . .	. . . .
Hydrogen	9.0×10^{-2}	14	20

The gas densities refer to a temperature of 273 K and a pressure of $1.01 \times 10^{5}\,\text{Pa}$.

SECTION A

For questions 1 to 20 in this section of the paper the answer to each question is either A, B, C, D or E. Decide what your answer is, then, using your pencil, put a horizontal line in the space provided—see the example below.

EXAMPLE

The energy unit measured by the electricity meter in your home is the

    A kilowatt-hour

    B ampere

    C watt

    D coulomb

    E volt.

The correct answer is **A**—kilowatt-hour. The answer **A** has been clearly marked in **pencil** with a horizontal line (see below).

Changing an answer

If you decide to change your answer, carefully erase your first answer and, using your pencil, fill in the answer you want. The answer below has been changed to **E**.

[**Turn over**

SECTION A

Answer questions 1–20 on the answer sheet.

1. Which of the following is a scalar quantity?

   A   velocity

   B   acceleration

   C   mass

   D   force

   E   momentum

2. A vehicle is travelling in a straight line. Graphs of velocity and acceleration against time are shown.

   Which pair of graphs could represent the motion of the vehicle?

   A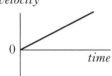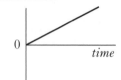

   B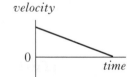

   C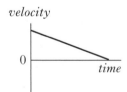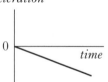

   D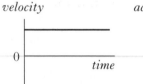

   E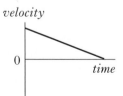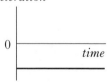

3. A car of mass 1200 kg pulls a horsebox of mass 700 kg along a straight, horizontal road.

   They have an acceleration of $2 \cdot 0 \, \text{m s}^{-2}$.

   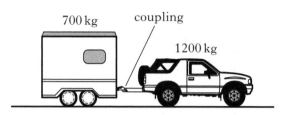

   Assuming that the frictional forces are negligible, the tension in the coupling between the car and the horsebox is

   A   500 N

   B   700 N

   C   1400 N

   D   2400 N

   E   3800 N.

4. Two trolleys travel towards each other in a straight line along a frictionless surface.

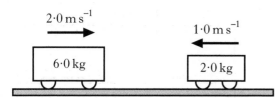

The trolleys collide. After the collision the trolleys move as shown below.

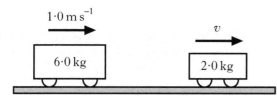

Which row in the table gives the total momentum and the total kinetic energy **after** the collision?

	Total momentum/ $\mathrm{kg\,m\,s^{-1}}$	Total kinetic energy/ J
A	10	7·0
B	10	13
C	10	20
D	14	13
E	14	7·0

5. An aircraft cruises at an altitude at which the external air pressure is $0\cdot40 \times 10^5\,\mathrm{Pa}$. The air pressure inside the aircraft cabin is maintained at $1\cdot0 \times 10^5\,\mathrm{Pa}$. The area of an external cabin door is $2\cdot0\,\mathrm{m}^2$.

What is the outward force on the door due to the pressure difference?

A $0\cdot30 \times 10^5\,\mathrm{N}$

B $0\cdot70 \times 10^5\,\mathrm{N}$

C $1\cdot2 \times 10^5\,\mathrm{N}$

D $2\cdot0 \times 10^5\,\mathrm{N}$

E $2\cdot8 \times 10^5\,\mathrm{N}$

6. A cylinder of height $1\cdot0\,\mathrm{m}$ is held stationary in a swimming pool. The top of the cylinder is at a depth of $1\cdot5\,\mathrm{m}$ below the surface of the water.

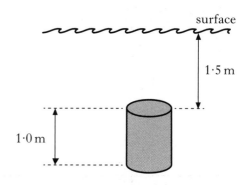

The density of the water is $1\cdot0 \times 10^3\,\mathrm{kg\,m^{-3}}$.

The pressure due to the water exerted on the top surface of the cylinder is

A $1\cdot5 \times 10^3\,\mathrm{N\,m^{-2}}$

B $4\cdot9 \times 10^3\,\mathrm{N\,m^{-2}}$

C $9\cdot8 \times 10^3\,\mathrm{N\,m^{-2}}$

D $14\cdot7 \times 10^3\,\mathrm{N\,m^{-2}}$

E $24\cdot5 \times 10^3\,\mathrm{N\,m^{-2}}$.

7. A fixed mass of gas is heated inside a rigid container. As its temperature changes from T_1 to T_2 the pressure increases from $1\cdot0 \times 10^5\,\mathrm{Pa}$ to $2\cdot0 \times 10^5\,\mathrm{Pa}$.

Which row in the table shows possible values for T_1 and T_2?

	T_1	T_2
A	27 °C	327 °C
B	30 °C	60 °C
C	80 °C	40 °C
D	303 K	333 K
E	600 K	300 K

[Turn over

8. One volt is equivalent to one

   A   farad per coulomb

   B   ampere per ohm

   C   joule per ampere

   D   joule per ohm

   E   joule per coulomb.

9. A Wheatstone bridge circuit is set up as shown.

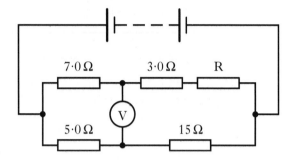

The reading on the voltmeter is zero.

The value of resistor R is

   A $3 \cdot 0\,\Omega$

   B $4 \cdot 0\,\Omega$

   C $18\,\Omega$

   D $21\,\Omega$

   E $24\,\Omega$.

10. A Wheatstone bridge circuit is set up as shown.

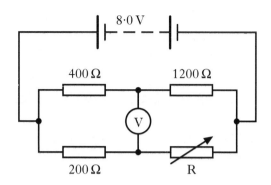

When the variable resistor R is set at $600\,\Omega$ the bridge is balanced. When R is set at $601\,\Omega$ the reading on the voltmeter is $+2 \cdot 5\,mV$.

R is now set at $598\,\Omega$.

The reading on the voltmeter is

   A $-7 \cdot 5\,mV$

   B $-5 \cdot 0\,mV$

   C $-2 \cdot 5\,mV$

   D $+5 \cdot 0\,mV$

   E $+7 \cdot 5\,mV$.

11. The output of a 50 Hz a.c. supply is connected to the input of an oscilloscope. The trace produced on the screen of the oscilloscope is shown.

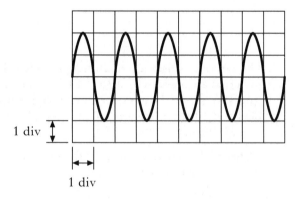

The time-base control of the oscilloscope is set at

   A   1 ms/div

   B   10 ms/div

   C   20 ms/div

   D   100 ms/div

   E   200 ms/div.

12. An a.c. supply with an output voltage of 6·0 V r.m.s. is connected to a 3·0 Ω resistor.

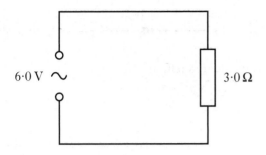

Which row in the table shows the peak voltage across the resistor and the peak current in the circuit?

	Peak voltage/V	Peak current/A
A	$6\sqrt{2}$	$2\sqrt{2}$
B	$6\sqrt{2}$	2
C	6	2
D	$\dfrac{6}{\sqrt{2}}$	$\dfrac{2}{\sqrt{2}}$
E	6	$2\sqrt{2}$

13. In an experiment to find the capacitance of a capacitor, a student makes the following measurements.

potential difference across capacitor $= (10\cdot0 \pm 0\cdot1)\,\mathrm{V}$

charge stored by capacitor $= (500 \pm 25)\,\mu\mathrm{C}$

Which row in the table gives the capacitance of the capacitor and the percentage uncertainty in the capacitance?

	Capacitance/μF	Percentage uncertainty
A	0·02	1
B	0·02	5
C	50	1
D	50	5
E	5000	6

14. A capacitor is connected to an a.c. supply and a.c. ammeter as shown.

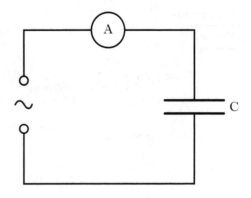

The supply has a constant peak voltage, but its frequency can be varied.

Which graph shows how the current I varies with the frequency f of the supply?

A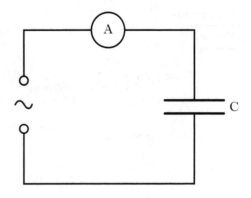

B

C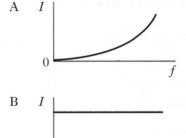

D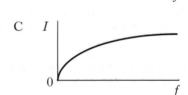

E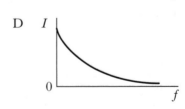

[Turn over

15. Two identical loudspeakers, L_1 and L_2, are connected to a signal generator as shown.

signal generator

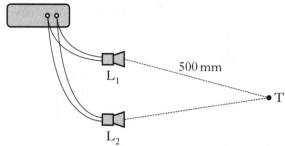

An interference pattern is produced.

A minimum is detected at point T.

The wavelength of the sound is 40 mm.

The distance from L_1 to T is 500 mm.

The distance from L_2 to T is

A   450 mm

B   460 mm

C   470 mm

D   480 mm

E   490 mm.

16. The irradiance of light from a point source is $20\,\text{W}\,\text{m}^{-2}$ at a distance of $5\cdot0\,\text{m}$ from the source.

What is the irradiance of the light at a distance of 25 m from the source?

A $0\cdot032\,\text{W}\,\text{m}^{-2}$

B $0\cdot80\,\text{W}\,\text{m}^{-2}$

C $4\cdot0\,\text{W}\,\text{m}^{-2}$

D $100\,\text{W}\,\text{m}^{-2}$

E $500\,\text{W}\,\text{m}^{-2}$

17. The diagram below represents part of the process of stimulated emission in a laser.

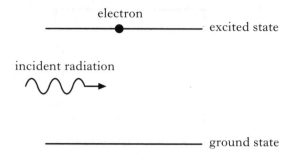

Which statement best describes the emitted radiation?

A   Out of phase and emitted in the same direction as the incident radiation.

B   Out of phase and emitted in the opposite direction to the incident radiation.

C   Out of phase and emitted in all directions.

D   In phase and emitted in the same direction as the incident radiation.

E   In phase and emitted in the opposite direction to the incident radiation.

18. In an n-type semiconductor

A   the majority charge carriers are electrons

B   the majority charge carriers are holes

C   the majority charge carriers are protons

D   there are more protons than electrons

E   there are more electrons than protons.

19. The following statement represents a nuclear decay.

$$^{214}_{x}\text{Pb} \rightarrow ^{y}_{83}\text{Bi} + ^{o}_{z}\text{e}$$

Which row in the table shows the correct values of x, y and z for this decay?

	x	y	z
A	82	210	−1
B	82	214	−1
C	84	214	1
D	85	210	2
E	85	214	2

20. The graph shows how the corrected count rate from a radioactive source varies with the thickness of a material placed between the source and a detector.

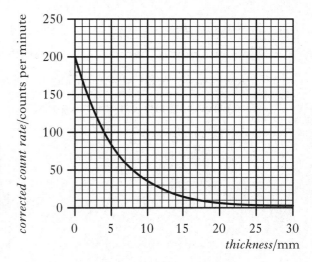

The half value thickness of the material is

A   4 mm

B   14 mm

C   28 mm

D   100 mm

E   200 mm.

[Turn over

SECTION B

Marks

Write your answers to questions 21 to 30 in the answer book.

21. A student investigates the motion of a ball projected from a launcher.

   The launcher is placed on the ground and a ball is fired vertically upwards.

   The vertical speed of the ball as it leaves the top of the launcher is $7 \cdot 0 \, \text{m s}^{-1}$.

   The effects of air resistance can be ignored.

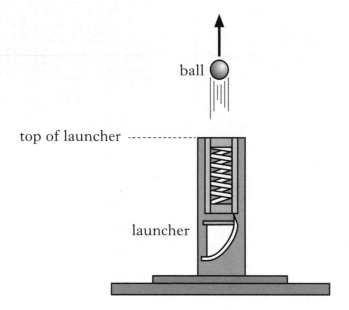

   (a) (i) Calculate the maximum height above the top of the launcher reached by the ball.   **2**

   (ii) Show that the time taken for the ball to reach its maximum height is $0 \cdot 71 \, \text{s}$.   **1**

Marks

21. (continued)

(b) The student now fixes the launcher to a trolley. The trolley travels horizontally at a constant speed of $1 \cdot 5\,\mathrm{m\,s^{-1}}$ to the right.

The launcher again fires the ball vertically upwards with a speed of $7 \cdot 0\,\mathrm{m\,s^{-1}}$.

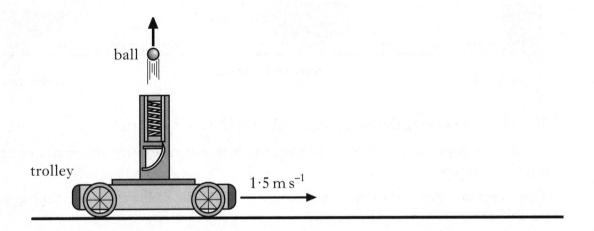

  (i) Determine the velocity of the ball after $0 \cdot 71\,\mathrm{s}$.   **1**

  (ii) The student asks some friends to predict where the ball will land relative to the moving launcher. They make the following statements.

Statement X: *The ball will land behind the launcher.*

Statement Y: *The ball will land in front of the launcher.*

Statement Z: *The ball will land on top of the launcher.*

Which of the statements is correct?

You must justify your answer.   **2**

(6)

[Turn over

22. An experiment is set up to investigate the motion of a cart as it collides with a force sensor.

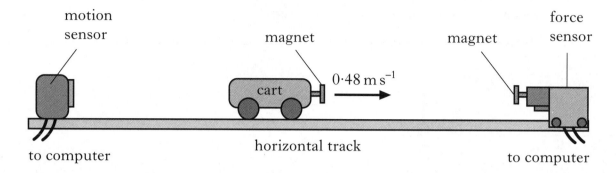

The cart moves along the horizontal track at $0.48\,\text{m s}^{-1}$ to the right.

As the cart approaches the force sensor, the magnets repel each other and exert a force on the cart.

The computer attached to the force sensor displays the following force-time graph for this collision.

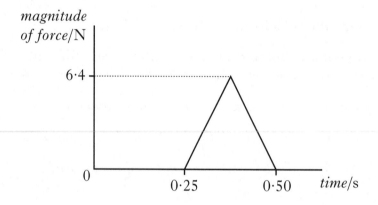

The computer attached to the motion sensor displays the following velocity-time graph for the cart.

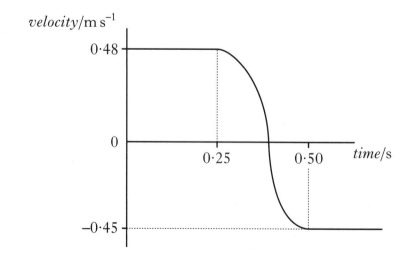

Marks

22. **(continued)**

(*a*)   (i) Calculate the magnitude of the impulse on the cart during the collision.   **2**

(ii) Determine the magnitude and direction of the change in momentum of the cart.   **1**

(iii) Calculate the mass of the cart.   **2**

(*b*) The experiment is repeated using different magnets which produce a greater average force on the cart during the collision. As before, the cart is initially travelling at $0.48 \, \text{m s}^{-1}$ to the right and the collision causes the same change in its velocity.

Copy the force-time graph shown and, on the same axes, draw another graph to show how the magnitude of the force varies with time in this collision.

Numerical values are not required but you must label each graph clearly.   **2**

(7)

[Turn over

Marks

23. A technician uses the equipment shown to calculate a value for the density of air at room temperature.

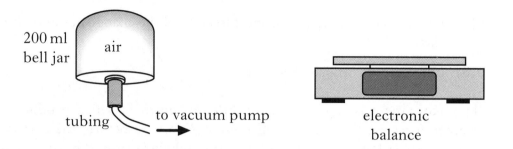

The mass of the bell jar is measured when it is full of air. The vacuum pump is then used to remove air from the bell jar. The mass of the bell jar is measured again.

The following measurements are obtained.

Mass before air is removed $= 111 \cdot 49 \, g$

Mass after air is removed $= 111 \cdot 26 \, g$

Volume of bell jar $= 200 \, ml = 2 \cdot 0 \times 10^{-4} \, m^3$

(a) (i) Use these measurements to calculate a value for the density of air in $kg \, m^{-3}$.   **2**

(ii) The accepted value for the density of air at this temperature is $1 \cdot 29 \, kg \, m^{-3}$. Explain why the technician's answer is different from the accepted value.   **1**

(b) Air is allowed back into the bell jar until it reaches a pressure of $1 \cdot 01 \times 10^5 \, Pa$.

The technician now uses a syringe to remove 50 ml of the air from the bell jar.

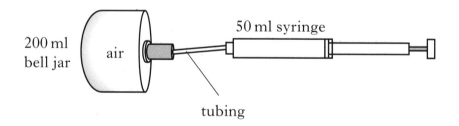

The temperature of the air remains constant.

(i) Calculate the new pressure of the air inside the bell jar.   **2**

(ii) Use the kinetic model to explain this change in pressure after removing air with the syringe.   **2**

(7)

Marks

24. (*a*) A supply of e.m.f. 10·0 V and internal resistance *r* is connected in a circuit as shown in Figure 1.

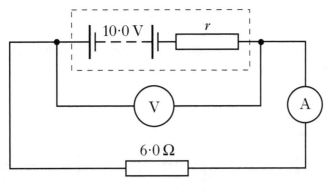

Figure 1

The meters display the following readings.

Reading on ammeter = 1·25 A

Reading on voltmeter = 7·50 V

   (i) What is meant by an *e.m.f. of 10·0 V*?   **1**

   (ii) Show that the internal resistance, *r*, of the supply is 2·0 Ω.   **1**

(*b*) A resistor R is connected to the circuit as shown in Figure 2.

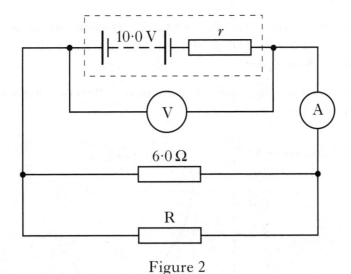

Figure 2

The meters now display the following readings.

Reading on ammeter = 2·0 A

Reading on voltmeter = 6·0 V

   (i) Explain why the reading on the voltmeter has decreased.   **2**

   (ii) Calculate the resistance of resistor R.   **3**

                (7)

Marks

25. A student carries out an experiment using a circuit which includes a capacitor with a capacitance of $200\,\mu F$.

(a) Explain what is meant by a *capacitance of 200 μF*.

1

(b) The capacitor is used in the circuit shown to measure the time taken for a ball to fall vertically between two strips of metal foil.

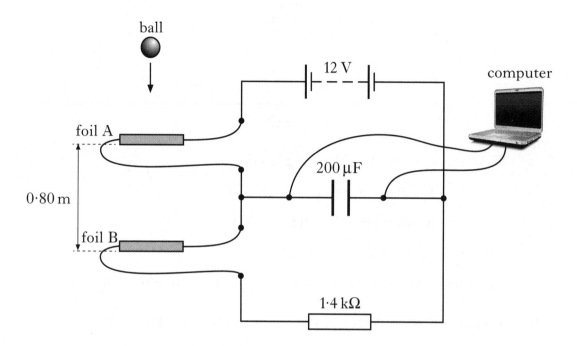

The ball is dropped from rest above foil A. It is travelling at $1\cdot5\,\mathrm{m\,s^{-1}}$ when it reaches foil A. It breaks foil A, then a short time later breaks foil B. These strips of foil are $0\cdot80\,\mathrm{m}$ apart.

The computer displays a graph of potential difference across the capacitor against time as shown.

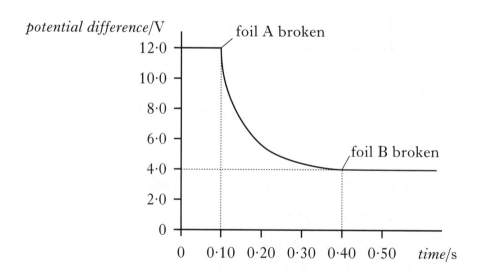

(i) Calculate the current in the $1\cdot4\,\mathrm{k\Omega}$ resistor at the moment foil A is broken.

2

(ii) Calculate the **decrease** in the energy stored in the capacitor during the time taken for the ball to fall from foil A to foil B.

3

Marks

25. (continued)

(*c*) The measurements from this experiment are now used to estimate the acceleration due to gravity.

    (i) What is the time taken for the ball to fall from foil A to foil B?   **1**

    (ii) Use the results of this experiment to calculate a value for the acceleration due to gravity.   **2**

    (iii) The distance between the two foils is now increased and the experiment repeated. Explain why this gives a more accurate result for the acceleration due to gravity.   **1**

(10)

[Turn over

Marks

26. (a) An op-amp is connected in a circuit as shown.

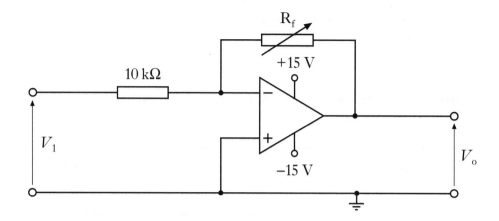

The resistance of the feedback resistor R_f is varied between $20\,k\Omega$ and $120\,k\Omega$.

The graph shows how the output voltage V_o varies as the resistance of the feedback resistor is increased.

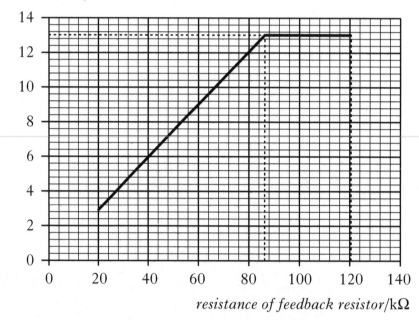

output voltage V_o/V

resistance of feedback resistor/kΩ

    (i)   In which mode is the op-amp being used?   1

   (ii)   Calculate the input voltage V_1.   2

  (iii)   Explain why the output voltage V_o does not increase above 13 V.   1

Marks

26. (continued)

(*b*) The op-amp is now connected in a different circuit as shown.

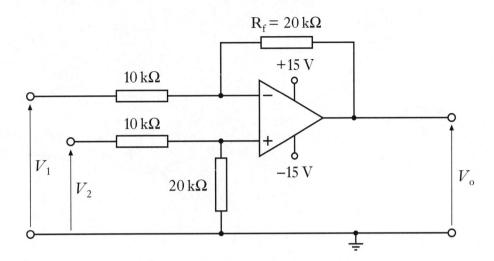

The input voltages V_1 and V_2 are now varied and the corresponding output voltage V_o is measured.

Graph 1 shows the input voltage V_1 for the first 3 seconds.

Graph 2 shows the input voltage V_2 for the first 3 seconds.

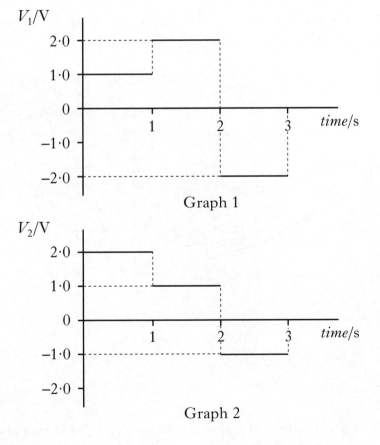

Graph 1

Graph 2

Sketch a graph to show the output voltage V_o from the op-amp for the first 3 seconds.

Numerical values are required on both the voltage and time axes.   **3**

(7)

Marks

27. (*a*) A ray of red light is incident on a block of glass as shown.

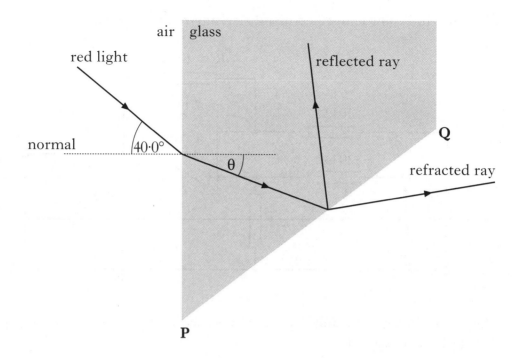

The refractive index of the glass for this light is 1·66.

  (i) Calculate the value of the angle θ shown in the diagram.   2

  (ii) The direction of the incident light ray is now changed so that the refracted ray emerges along face **PQ** as shown.

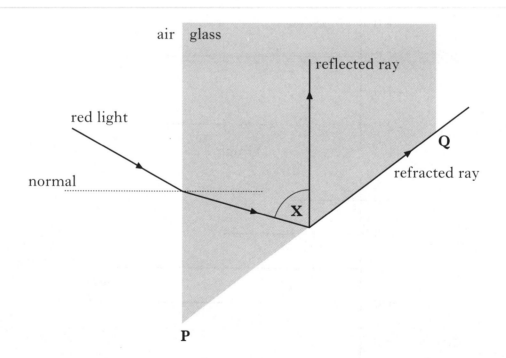

    (A) Calculate the critical angle for the red light in this glass.   2

    (B) Determine the size of angle **X** shown in the diagram.   1

Marks

27. (continued)

(b) The ray of red light is now replaced with a ray of blue light.

This ray of blue light is directed towards the block along the same path as the ray of red light in part (a)(ii).

Is this ray of blue light refracted at face **PQ**?

Justify your answer.   2

(7)

[Turn over

Marks

28. (*a*) The first demonstration of the interference of light was performed by Thomas Young in 1801.

What does the demonstration of interference prove about light?   **1**

(*b*) A grating is placed in a colourless liquid in a container. Laser light is incident on the grating along the normal. The spacing between the lines on the grating is $5 \cdot 0 \times 10^{-6}$ m. Interference occurs and the maxima produced are shown in the diagram.

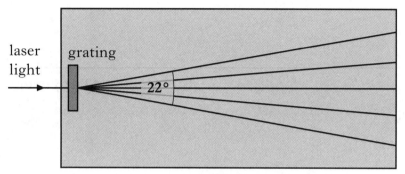

container filled with a colourless liquid

(i) Calculate the wavelength of the laser light in the liquid.   **2**

(ii) The refractive index of the colourless liquid decreases as the temperature of the liquid increases.

The liquid is now heated.

What effect does this have on the spacing between the maxima?

You must justify your answer.   **2**

(5)

29. A metal plate emits electrons when certain wavelengths of electromagnetic radiation are incident on it.

Marks

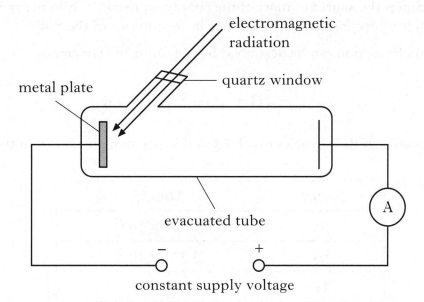

The work function of the metal is $2 \cdot 24 \times 10^{-19}$ J.

(a) Electrons are released when electromagnetic radiation of wavelength 525 nm is incident on the surface of the metal plate.

    (i) Show that the energy of each photon of the incident radiation is $3 \cdot 79 \times 10^{-19}$ J.

    2

    (ii) Calculate the maximum kinetic energy of an electron released from the surface of the metal plate.

    1

(b) The frequency of the incident radiation is now varied through a range of values.

The maximum kinetic energy of electrons leaving the metal plate is determined for each frequency.

A graph of this maximum kinetic energy against frequency is shown.

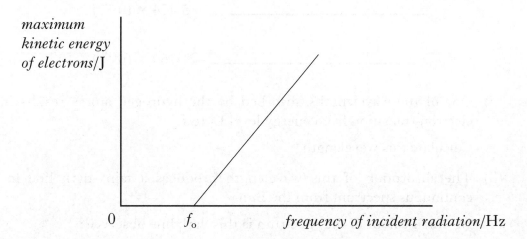

    (i) Explain why the kinetic energy of the electrons is zero below the frequency f_o.

    1

    (ii) Calculate the value of the frequency f_o.

    2

    (6)

[Turn over for Question 30 on *Page twenty-four*

Marks

30. (*a*) The Sun is the source of most of the energy on Earth. This energy is produced by nuclear reactions which take place in the interior of the Sun.

One such reaction can be described by the following statement.

$$^{3}_{1}H + {}^{2}_{1}H \rightarrow {}^{4}_{2}He + {}^{1}_{0}n$$

The masses of the particles involved in this reaction are shown in the table.

Particle	Mass/kg
$^{3}_{1}H$	$5{\cdot}005 \times 10^{-27}$
$^{2}_{1}H$	$3{\cdot}342 \times 10^{-27}$
$^{4}_{2}He$	$6{\cdot}642 \times 10^{-27}$
$^{1}_{0}n$	$1{\cdot}675 \times 10^{-27}$

  (i) Name this type of nuclear reaction.   **1**

  (ii) Calculate the energy released in this reaction.   **3**

(*b*) The Sun emits a continuous spectrum of visible light. When this light passes through hydrogen atoms in the Sun's outer atmosphere, certain wavelengths are absorbed.

The diagram shows some of the energy levels for the hydrogen atom.

E_3 ———————————— $-1{\cdot}360 \times 10^{-19}$ J

E_2 ———————————— $-2{\cdot}416 \times 10^{-19}$ J

E_1 ———————————— $-5{\cdot}424 \times 10^{-19}$ J

E_0 ———————————— $-21{\cdot}760 \times 10^{-19}$ J

  (i) One of the wavelengths absorbed by the hydrogen atoms results in an electron transition from energy level E_1 to E_3.

Calculate this wavelength.   **3**

  (ii) The absorption of this wavelength produces a faint dark line in the continuous spectrum from the Sun.

In which colour of the spectrum is this dark line observed?   **1**

(8)

[END OF QUESTION PAPER]

[BLANK PAGE]

X069/12/02

NATIONAL QUALIFICATIONS 2012	MONDAY, 28 MAY 1.00 PM – 3.30 PM	PHYSICS HIGHER

Read Carefully

Reference may be made to the Physics Data Booklet.

1 All questions should be attempted.

Section A (questions 1 to 20)

2 Check that the answer sheet is for Physics Higher (Section A).

3 For this section of the examination you must use an **HB pencil** and, where necessary, an eraser.

4 Check that the answer sheet you have been given has **your name**, **date of birth**, **SCN** (Scottish Candidate Number) and **Centre Name** printed on it.

 Do not change any of these details.

5 If any of this information is wrong, tell the Invigilator immediately.

6 If this information is correct, **print** your name and seat number in the boxes provided.

7 There is **only one correct** answer to each question.

8 Any rough working should be done on the question paper or the rough working sheet, **not** on your answer sheet.

9 At the end of the exam, put the **answer sheet for Section A inside the front cover of your answer book**.

10 Instructions as to how to record your answers to questions 1–20 are given on page three.

Section B (questions 21 to 31)

11 Answer the questions numbered 21 to 31 in the answer book provided.

12 **All answers must be written clearly and legibly in ink**.

13 Fill in the details on the front of the answer book.

14 Enter the question number clearly in the margin of the answer book beside each of your answers to questions 21 to 31.

15 Care should be taken to give an appropriate number of significant figures in the final answers to calculations.

16 Where additional paper, eg square ruled paper, is used, write your name and SCN (Scottish Candidate Number) on it and place it inside the front cover of your answer booklet.

DATA SHEET
COMMON PHYSICAL QUANTITIES

Quantity	Symbol	Value	Quantity	Symbol	Value
Speed of light in vacuum	c	$3.00 \times 10^8\,\mathrm{m\,s^{-1}}$	Mass of electron	m_e	$9.11 \times 10^{-31}\,\mathrm{kg}$
Magnitude of the charge on an electron	e	$1.60 \times 10^{-19}\,\mathrm{C}$	Mass of neutron	m_n	$1.675 \times 10^{-27}\,\mathrm{kg}$
Gravitational acceleration on Earth	g	$9.8\,\mathrm{m\,s^{-2}}$	Mass of proton	m_p	$1.673 \times 10^{-27}\,\mathrm{kg}$
Planck's constant	h	$6.63 \times 10^{-34}\,\mathrm{J\,s}$			

REFRACTIVE INDICES

The refractive indices refer to sodium light of wavelength 589 nm and to substances at a temperature of 273 K.

Substance	Refractive index	Substance	Refractive index
Diamond	2·42	Water	1·33
Crown glass	1·50	Air	1·00

SPECTRAL LINES

Element	Wavelength/nm	Colour	Element	Wavelength/nm	Colour
Hydrogen	656	Red	Cadmium	644	Red
	486	Blue-green		509	Green
	434	Blue-violet		480	Blue
	410	Violet		*Lasers*	
	397	Ultraviolet	Element	Wavelength/nm	Colour
	389	Ultraviolet	Carbon dioxide	9550 ⎱ 10590 ⎰	Infrared
Sodium	589	Yellow	Helium-neon	633	Red

PROPERTIES OF SELECTED MATERIALS

Substance	Density/kg m⁻³	Melting Point/K	Boiling Point/K
Aluminium	2.70×10^3	933	2623
Copper	8.96×10^3	1357	2853
Ice	9.20×10^2	273	. . . .
Sea Water	1.02×10^3	264	377
Water	1.00×10^3	273	373
Air	1·29	. . . .	. . . .
Hydrogen	9.0×10^{-2}	14	20

The gas densities refer to a temperature of 273 K and a pressure of $1.01 \times 10^5\,\mathrm{Pa}$.

SECTION A

For questions 1 to 20 in this section of the paper the answer to each question is either A, B, C, D or E. Decide what your answer is, then, using your pencil, put a horizontal line in the space provided—see the example below.

EXAMPLE

The energy unit measured by the electricity meter in your home is the

  A kilowatt-hour

  B ampere

  C watt

  D coulomb

  E volt.

The correct answer is **A**—kilowatt-hour. The answer **A** has been clearly marked in **pencil** with a horizontal line (see below).

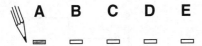

Changing an answer

If you decide to change your answer, carefully erase your first answer and, using your pencil, fill in the answer you want. The answer below has been changed to **E**.

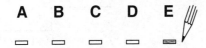

[Turn over

SECTION A

Answer questions 1–20 on the answer sheet.

1. Which of the following contains one vector and two scalar quantities?

   A   force, time and acceleration

   B   power, momentum and velocity

   C   acceleration, velocity and force

   D   mass, distance and speed

   E   acceleration, time and speed

2. A trolley travels along a straight track.

   The graph shows how the velocity v of the trolley varies with time t.

   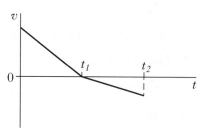

   Which graph shows how the acceleration a of the trolley varies with time t?

   A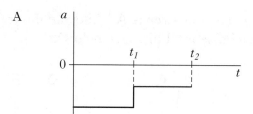

   B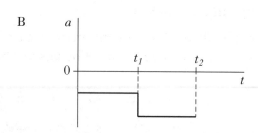

   C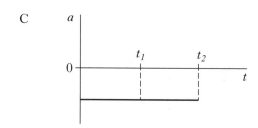

   D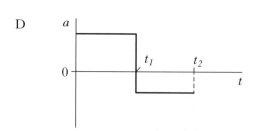

   E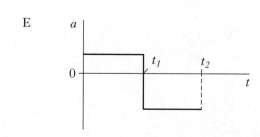

3. A rocket of mass 200 kg accelerates vertically upwards from the surface of a planet at $2 \cdot 0 \, \text{m s}^{-2}$.

The gravitational field strength on the planet is $4 \cdot 0 \, \text{N kg}^{-1}$.

What is the size of the force being exerted by the rocket's engines?

A   400 N

B   800 N

C   1200 N

D   2000 N

E   2400 N

4. The diagram shows the masses and velocities of two trolleys just before they collide on a level bench.

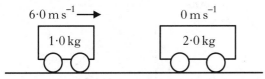

After the collision, the trolleys move along the bench joined together.

How much kinetic energy is lost in this collision?

A   0 J

B   6·0 J

C   12 J

D   18 J

E   24 J

5. A fixed mass of gas condenses at atmospheric pressure to form a liquid.

Which row in the table shows the approximate increase in density and the approximate decrease in spacing between molecules?

	Approximate increase in density	Approximate decrease in spacing between molecules
A	10 times	10 times
B	10 times	1000 times
C	1000 times	10 times
D	1000 times	1000 times
E	1 000 000 times	1000 times

6. Two identical blocks are suspended in water at different depths as shown.

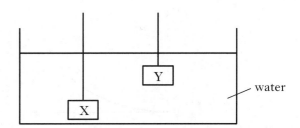

A student makes the following statements.

I   The buoyancy force on block Y is greater than the buoyancy force on block X.

II   The pressure on the bottom of block X is greater than the pressure on the bottom of block Y.

III   The pressure on the top of block X is greater than the pressure on the top of block Y.

Which of the statements is/are correct?

A   I only

B   II only

C   I and II only

D   II and III only

E   I, II and III

[Turn over

7. Which of the following graphs shows the relationship between the pressure P and the volume V of a fixed mass of gas at constant temperature?

A

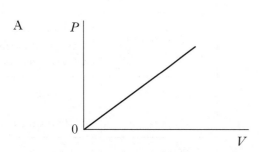

B

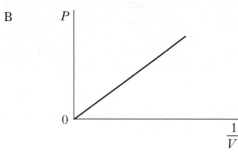

C

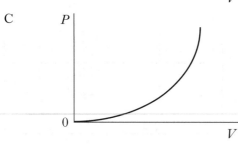

D

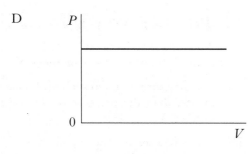

E

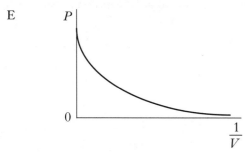

8. A circuit is set up as shown.

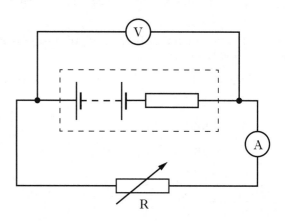

The variable resistor R is adjusted and a series of readings taken from the voltmeter and ammeter.

The graph shows how the voltmeter reading varies with the ammeter reading.

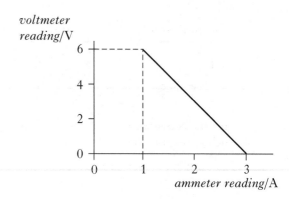

Which row in the table shows the values for the e.m.f. and internal resistance of the battery in the circuit?

	e.m.f./V	internal resistance/Ω
A	6	2
B	6	3
C	9	2
D	9	3
E	9	6

9. The diagram shows part of an electrical circuit.

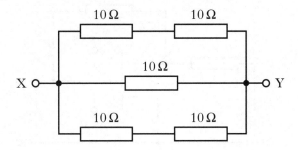

What is the resistance between X and Y?

A   0·2 Ω

B   5 Ω

C   10 Ω

D   20 Ω

E   50 Ω

10. An alternating voltage is displayed on an oscilloscope screen. The Y-gain and the timebase settings are shown.

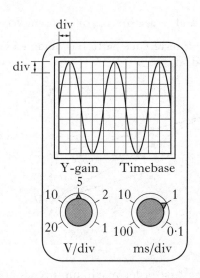

Which row in the table gives the values for the peak voltage and frequency of the signal?

	Peak voltage/V	Frequency/Hz
A	10	100
B	10	250
C	20	250
D	10	500
E	20	1000

11. A student carries out an experiment to find the capacitance of a capacitor. The charge on the capacitor is measured for different values of p.d. across the capacitor. The results are shown.

charge on capacitor/μC	p.d. across capacitor/V
1·9	1·0
4·6	2·0
9·6	4·0

The best estimate of the capacitance is

A   1·9 μF

B   2·2 μF

C   2·3 μF

D   2·4 μF

E   2·6 μF.

[Turn over

12. The circuits below have identical a.c. supplies which are set at a frequency of 200 Hz.

constant amplitude
variable frequency

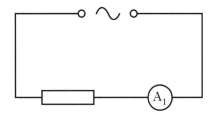

constant amplitude
variable frequency

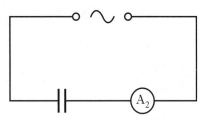

The frequency of each a.c. supply is now increased to 500 Hz.

What happens to the readings on the ammeters A_1 and A_2?

	A_1	A_2
A	increases	decreases
B	decreases	increases
C	no change	no change
D	no change	decreases
E	no change	increases

13. An op-amp circuit is set up as shown.

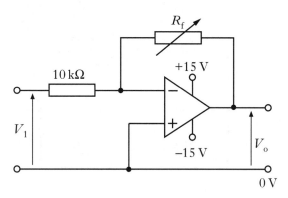

The resistance of R_f can be varied between 0 and 100 kΩ.

When the input voltage V_1 is +2 V a possible value of the output voltage V_o is

A   +20 V

B   +10 V

C   +2 V

D   −10 V

E   −20 V.

14. S_1 and S_2 are sources of coherent waves.

An interference pattern is obtained between X and Y.

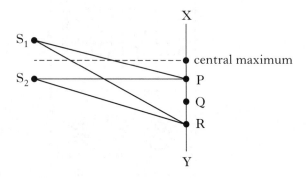

The first order maximum occurs at P, where $S_1P = 200$ mm and $S_2P = 180$ mm.

For the third order maximum, at R, the path difference $(S_1R - S_2R)$ is

A   20 mm

B   30 mm

C   40 mm

D   50 mm

E   60 mm.

15. Clean zinc plates are mounted on insulating handles and then charged.

Different types of electromagnetic radiation are now incident on the plates as shown.

Which of the zinc plates is most likely to discharge due to photoelectric emission?

A

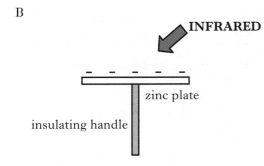

B

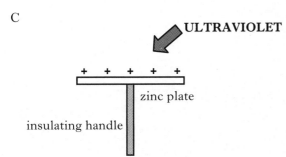

C

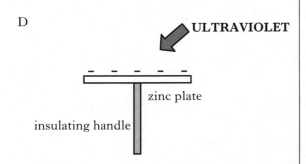

D

E

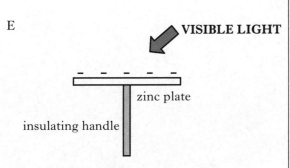

16. Electromagnetic radiation of frequency $9.0 \times 10^{14}\,\mathrm{Hz}$ is incident on a clean metal surface.

The work function of the metal is $5.0 \times 10^{-19}\,\mathrm{J}$.

The maximum kinetic energy of a photoelectron released from the metal surface is

A $1.0 \times 10^{-19}\,\mathrm{J}$

B $4.0 \times 10^{-19}\,\mathrm{J}$

C $5.0 \times 10^{-19}\,\mathrm{J}$

D $6.0 \times 10^{-19}\,\mathrm{J}$

E $9.0 \times 10^{-19}\,\mathrm{J}$.

17. In an atom, a photon of radiation is emitted when an electron makes a transition from a higher energy level to a lower energy level as shown.

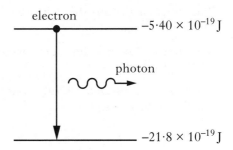

The wavelength of the radiation emitted due to an electron transition between the two energy levels shown is

A $1.2 \times 10^{-7}\,\mathrm{m}$

B $7.3 \times 10^{-8}\,\mathrm{m}$

C $8.2 \times 10^{6}\,\mathrm{m}$

D $1.4 \times 10^{7}\,\mathrm{m}$

E $2.5 \times 10^{15}\,\mathrm{m}$.

[Turn over

18. Which of the following statements describes a spontaneous nuclear fission reaction?

A $^{235}_{92}U + ^{1}_{0}n \rightarrow ^{144}_{56}Ba + ^{90}_{36}Kr + 2^{1}_{0}n$

B $^{7}_{3}Li + ^{1}_{1}H \rightarrow ^{4}_{2}He + ^{4}_{2}He$

C $^{3}_{1}H + ^{2}_{1}H \rightarrow ^{4}_{2}He + ^{1}_{0}n$

D $^{226}_{88}Ra \rightarrow ^{222}_{86}Rn + ^{4}_{2}He$

E $^{216}_{84}Po \rightarrow ^{216}_{84}Po + \gamma$

19. The statement below represents a nuclear reaction.

$$^{3}_{1}H + ^{2}_{1}H \rightarrow ^{4}_{2}He + ^{1}_{0}n$$

The total mass on the left hand side is 8.347×10^{-27} kg.

The total mass on the right hand side is 8.316×10^{-27} kg.

The energy released during one nuclear reaction of this type is

A 9.30×10^{-21} J

B 2.79×10^{-12} J

C 7.51×10^{-10} J

D 1.50×10^{-9} J

E 2.79×10^{15} J.

20. A source of gamma radiation is stored in a large container. A count rate of 160 counts per minute, after correction for background radiation, is recorded outside the container.

The container is to be shielded so that the corrected count rate at the same point outside the container is no more than 10 counts per minute.

Lead and water are available as shielding materials. For this source, the half-value thickness of lead is 11 mm and the half-value thickness of water is 110 mm.

Which of the following shielding arrangements meets the above requirements?

A   40 mm of lead only

B   33 mm of lead plus 110 mm of water

C   20 mm of lead plus 220 mm of water

D   11 mm of lead plus 275 mm of water

E   10 mm of lead plus 330 mm of water

SECTION B

Write your answers to questions 21 to 31 in the answer book.

21. Two cyclists choose different routes to travel from point **A** to a point **B** some distance away.

(*a*) Cyclist X travels 12 km due East (bearing 090). He then turns onto a bearing of 200 (20 ° West of South) and travels a further 15 km to arrive at **B**. He takes 1 hour 15 minutes to travel from **A** to **B**.

   (i) By scale drawing (or otherwise) find the displacement of **B** from **A**.   **2**

   (ii) Calculate the average velocity of cyclist X for the journey from **A** to **B**.   **2**

(*b*) Cyclist Y travels a total distance of 33 km by following a different route from **A** to **B** at an average speed of 22 km h⁻¹.

   (i) State the displacement of cyclist Y on completing this route.   **1**

   (ii) Calculate the average velocity of cyclist Y for the journey from **A** to **B**.   **3**

                   (8)

[Turn over

Marks

22. A golfer hits a ball from point **P**. The ball leaves the club with a velocity v at an angle of θ to the horizontal.

   The ball travels through the air and lands at point **R**.

   Midway between **P** and **R** there is a tree of height 10·0 m.

not to scale

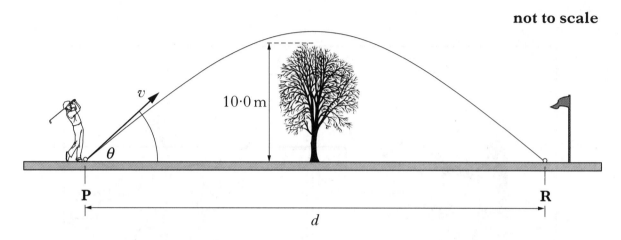

(a) The horizontal and vertical components of the ball's velocity during its flight are shown.

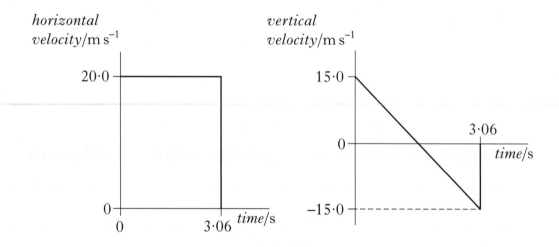

   The effects of air resistance can be ignored.

   Calculate:

   (i)   the horizontal distance d;   1

   (ii) the maximum height of the ball above the ground.   2

(b) When the effects of air resistance are **not** ignored, the golf ball follows a different path.

   Is the ball more or less likely to hit the tree?

   You must justify your answer.   2

                                                                                        (5)

Marks

23. An ion propulsion engine can be used to propel spacecraft to areas of deep space.

A simplified diagram of a Xenon ion engine is shown.

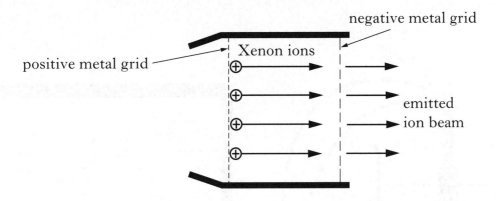

The Xenon ions are accelerated as they pass through an electric field between the charged metal grids. The emitted ion beam causes a force on the spacecraft in the opposite direction.

The spacecraft has a total mass of 750 kg.

The mass of a Xenon ion is $2 \cdot 18 \times 10^{-25}$ kg and its charge is $1 \cdot 60 \times 10^{-19}$ C. The potential difference between the charged metal grids is $1 \cdot 22$ kV.

(a) (i) Show that the work done on a Xenon ion as it moves through the electric field is $1 \cdot 95 \times 10^{-16}$ J.   1

(ii) Assuming the ions are accelerated from rest, calculate the speed of a Xenon ion as it leaves the engine.   2

(b) The ion beam exerts a constant force of $0 \cdot 070$ N on the spacecraft. Calculate the change in speed of the spacecraft during a 60 second period of time.   2

(c) A different ion propulsion engine uses Krypton ions which have a smaller mass than Xenon ions. The Krypton engine emits the same number of ions per second at the same speed as the Xenon engine.

Which of the two engines produces a greater force?

Justify your answer.   2

(7)

[Turn over

Marks

24. A student carries out an experiment to investigate the relationship between the pressure and temperature of a fixed mass of gas. The apparatus used is shown.

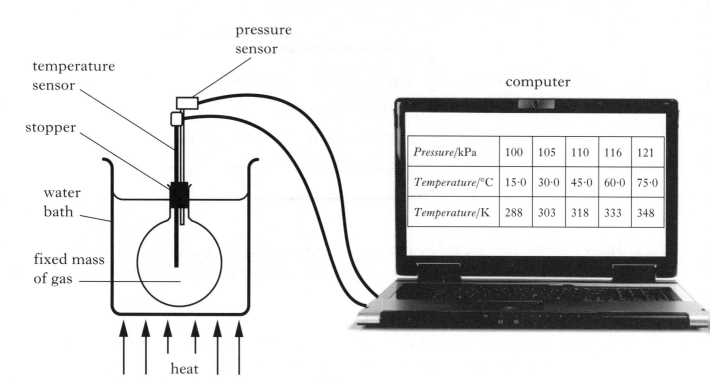

The pressure and temperature of the gas are recorded using sensors connected to a computer. The gas is heated slowly in the water bath and a series of readings is taken.

The volume of the gas remains constant during the experiment.

The results are shown.

Pressure/kPa	100	105	110	116	121
Temperature/°C	15·0	30·0	45·0	60·0	75·0
Temperature/K	288	303	318	333	348

(a) Using **all** the relevant data, establish the relationship between the pressure and the temperature of the gas.   2

(b) Use the kinetic model to explain the change in pressure as the temperature of the gas increases.   2

(c) Explain why the level of water in the water bath should be above the bottom of the stopper.   1

(5)

Marks

25. A student carries out two experiments using different power supplies connected to a lamp of resistance $6 \cdot 0 \, \Omega$.

(a) In the first experiment the lamp is connected to a power supply of e.m.f. 12 V and internal resistance $2 \cdot 0 \, \Omega$ as shown.

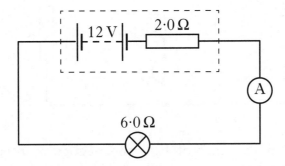

Calculate:

   (i) the reading on the ammeter;   2

  (ii) the lost volts;   1

 (iii) the output power of the lamp.   2

(b) In the second experiment the lamp is connected to a different power supply. This supply has the same e.m.f. as the supply in part (a) but a different value of internal resistance.

The output power of the lamp is now greater.

Assuming the resistance of the lamp has not changed, is the internal resistance of the new power supply less than, equal to, or greater than the internal resistance of the original supply?

Justify your answer.   2

(7)

[Turn over

Marks

26. The charging and discharging of a capacitor are investigated using the circuit shown.

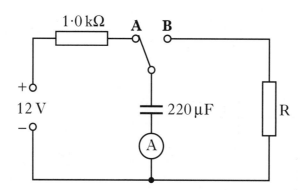

The power supply has an e.m.f. of 12 V and negligible internal resistance. The capacitor is initially uncharged.

(a) The switch is connected to **A** and the capacitor starts to charge. Sketch a graph showing how the voltage across the plates of the capacitor varies with time. Your graph should start from the moment the switch is connected to **A** until the capacitor is fully charged.

Numerical values are only required on the voltage axis. 2

(b) The capacitor is now discharged by moving the switch to **B**.

The graph of current against time as the capacitor discharges is shown.

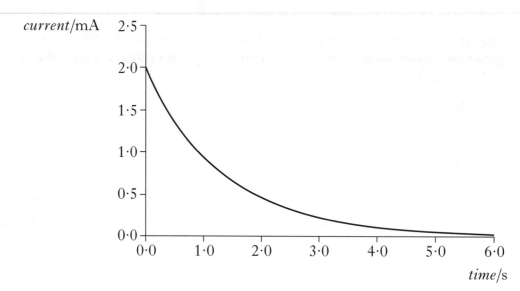

Calculate the resistance of R. 2

26.   (continued)   *Marks*

(c)   The 220 µF capacitor is now replaced with one of different value. This new capacitor is fully charged by moving the switch to **A**. It is then discharged by moving the switch to **B**.

The graph of current against time as this capacitor discharges is shown.

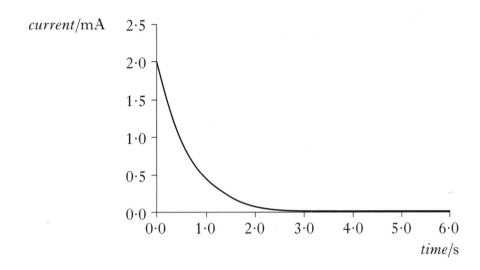

(i)   Explain why the value of the initial discharging current remains the same as in part (*b*).   1

(ii) How does the capacitance of this capacitor compare with the capacitance of the original 220 µF capacitor?

You must justify your answer.   2

                                                                          (7)

                                                                   [Turn over

[BLANK PAGE]

Marks

27. A fabric has been developed for use as a sensor in a breathing rate monitor. The graph shows how the resistance of a 50 mm length of this fabric changes as it is stretched.

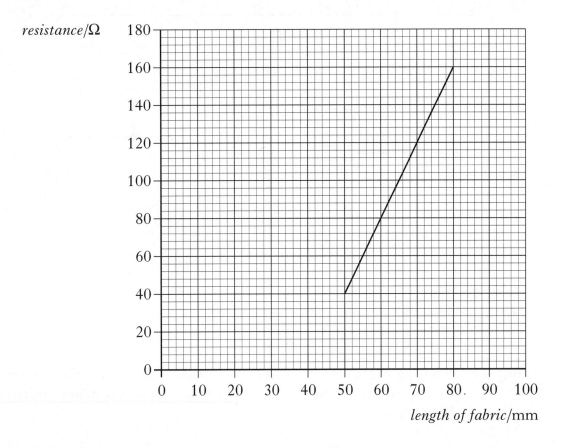

A sample of the fabric of unstretched length 50 mm is connected in a Wheatstone bridge circuit.

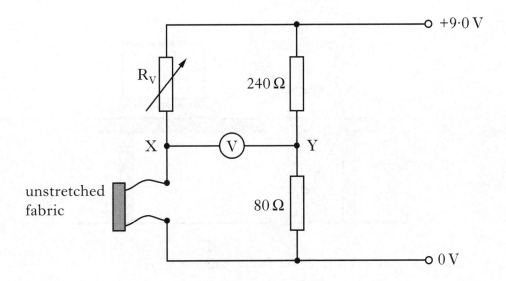

(a) The variable resistor R_V is adjusted until the bridge is balanced.

Show that the resistance of R_V is now 120 Ω.   2

27. (continued)   *Marks*

(b) The Wheatstone bridge is now connected to an op-amp circuit as shown.

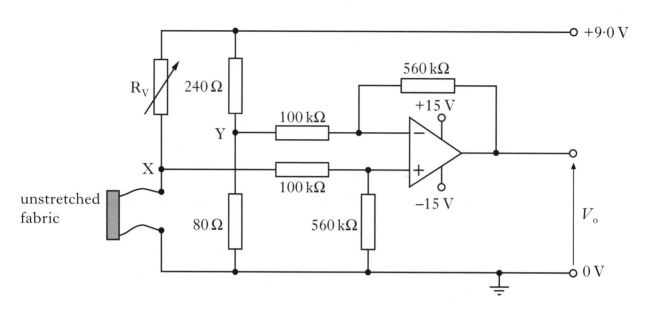

(i) In which mode is the op-amp being used?   1

(ii) Calculate the gain of the op-amp.   1

(iii) The 50 mm length of fabric remains connected in the circuit. This sensor is attached to a patient to monitor his chest movements. The fabric stretches as he breathes in.

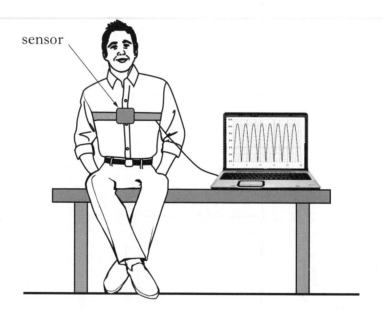

The potential at Y is 2·25 V. R_V remains at 120 Ω.

The output from the op-amp is connected to a computer.

The voltage V_o produced as the patient breathes in and out for 24 seconds is shown.

27. (*b*) (**iii**) (**continued**)   *Marks*

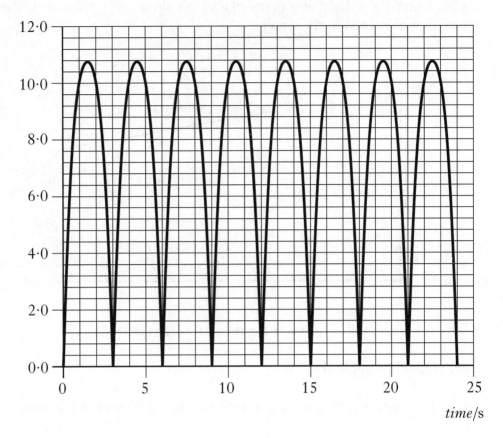

(A) Calculate the maximum potential difference between X and Y during this time.   **2**

(B) Calculate the maximum length of the fabric during this time.   **3**

(9)

[Turn over

Marks

28. A technician investigates the path of laser light as it passes through a glass tank filled with water. The light enters the glass tank along the normal at **C** then reflects off a mirror submerged in the water.

not to scale

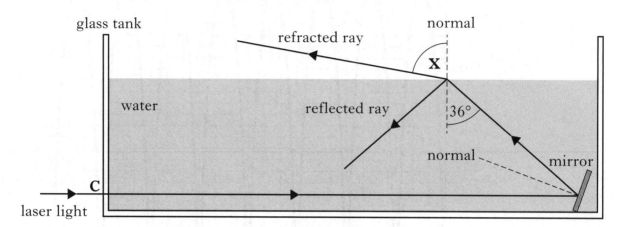

The refractive index of water for this laser light is 1·33.

(a) Calculate angle **X**.   2

(b) The mirror is now adjusted until the light follows the paths shown.

not to scale

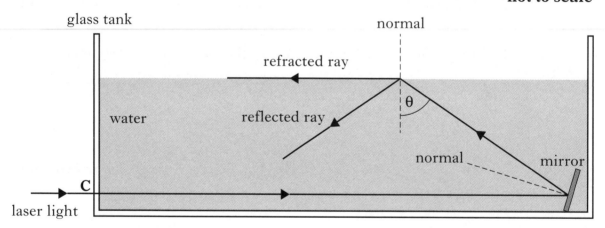

   (i) State why the value of θ is equal to the critical angle for this laser light in water.   1

   (ii) Calculate angle θ.   2

(c) The water is now replaced with a liquid which has a greater refractive index. The mirror is kept at the same angle as in part (b) and the incident ray again enters the tank along the normal at **C**.

   Draw a sketch which shows the path of the light ray after it has reflected off the mirror.

   Your sketch should only show what happens at the surface of the liquid.   1

(6)

Marks

29. A manufacturer claims that a grating consists of $3{\cdot}00 \times 10^5$ lines per metre and is accurate to $\pm\,2{\cdot}0\%$. A technician decides to test this claim. She directs laser light of wavelength 633 nm onto the grating.

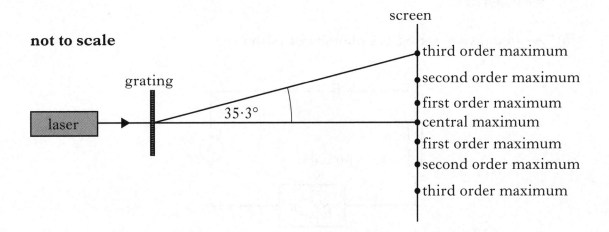

She measures the angle between the central maximum and the third order maximum to be $35{\cdot}3°$.

(a) Calculate the value she obtains for the slit separation for this grating.   2

(b) What value does she determine for the number of lines per metre for this grating?   1

(c) Does the technician's value for the number of lines per metre agree with the manufacturer's claim of $3{\cdot}00 \times 10^5$ lines per metre $\pm\,2{\cdot}0\%$?

You must justify your answer by calculation.   2

(5)

[Turn over

Marks

30. (*a*) An n-type semiconductor is formed by adding impurity atoms to a sample of pure semiconductor material.

State the effect that the addition of the impurity atoms has on the resistance of the material.

1

(*b*) A p-n junction is used as a photodiode as shown.

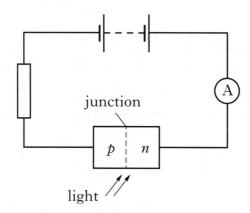

(i) In which mode is the photodiode operating?

1

(ii) The irradiance of the light on the junction of the photodiode is now increased.

Explain what happens to the current in the circuit.

2

(*c*) The photodiode is placed at a distance of 1·2 m from a small lamp. The reading on the ammeter is 3·0 μA.

The photodiode is now moved to a distance of 0·80 m from the same lamp. Calculate the new reading on the ammeter.

2

(6)

Marks

31. A medical physicist is investigating the effects of radiation on tissue samples.

One sample of tissue receives an absorbed dose of 500 μGy of radiation from a source.

The radiation weighting factors of different types of radiation are shown.

Type of radiation	*Radiation weighting factor (w_R)*
gamma	1
thermal neutrons	3
fast neutrons	10
alpha	20

(*a*) The tissue sample has a mass of 0·040 kg. Calculate the total energy absorbed by the tissue. **2**

(*b*) The average equivalent dose rate for this tissue sample is 5·00 mSv h^{-1}.
The tissue is exposed to radiation for 2 hours.

Which type of radiation is the medical physicist using?

Justify your answer by calculation. **3**

(5)

[*END OF QUESTION PAPER*]

Acknowledgements

Permission has been sought from all relevant copyright holders and Bright Red Publishing is grateful for the use of the following:

A picture of electronically heated gloves. Reproduced with permission of Zanier Sport (2008 page 15).

HIGHER | ANSWER SECTION

SQA HIGHER PHYSICS
2008–2012

SECTION A

1. D	2. A	3. B	4. C
5. D	6. C	7. D	8. E
9. E	10. D	11. A	12. B
13. E	14. C	15. A	16. B
17. B	18. C	19. C	20. D

SECTION B

21. (a) $v^2 = u^2 + 2as$

$12^2 = 30^2 + (2 \times -9 \times s)$

$s = 42$ m

(b) Speed at Q is greater/faster because:
- Deceleration/acceleration is less
- Mass of car is greater/bigger
- $a = F/m$ (and F is constant)

(c) (i) electrons and holes recombine at/in the junction (and energy is released)

(ii) $V_r = 12 - 5 = 7$ V

$I = \dfrac{P}{V}$

$= \dfrac{2 \cdot 2}{5} = 0 \cdot 44$ (A)

$R = \dfrac{V}{I}$

$= \dfrac{7}{0 \cdot 44}$

$= 16 \ \Omega$

22. (a) (i) $F = mg \sin\theta$

$= 40 \times 9 \cdot 8 \times \sin 30$

$= 196$ N

(ii) Balanced forces

or

$F = mg \sin\theta + $ Frictional force

$240 \quad = 196 + F_f$

$F_f \quad = 44$ N

(b) (i) Constant deceleration (of 6 m s^{-2})

or Constant acceleration (of -6 m s^{-2})

(ii)

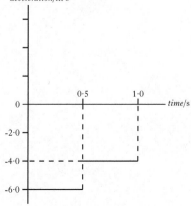

(iii) When crate moving up the slope, mgsinθ/comp of weight and friction are in the same direction. Moving back down the slope, forces are in opposite direction or friction has changed direction.

23. (a) (i) $\dfrac{P_1}{T_1} = \dfrac{P_2}{T_2}$

$\dfrac{2 \cdot 82 \times 10^6}{(19 + 273)} = \dfrac{P_2}{(5 + 273)}$

$P_2 = 2 \cdot 68 \times 10^6$ Pa

(ii) No change as both mass <u>and</u> volume remain constant and density = mass/volume ($\rho = m/V$)

(b) (i) $m = \rho V$

$m = 37 \cdot 6 \times 0 \cdot 03$

$m = 1 \cdot 13$ kg

(ii) Fewer molecules/atoms/particles inside canister so fewer collisions/hits **with walls** per second.

(iii) (gas stops escaping when) pressure inside = pressure outside

or

gas has reached atmospheric pressure

or

because $1 \cdot 01 \times 10^5$ Pa = atmospheric pressure

24. (a) (i) $4 \ \Omega$

(ii) $I = \dfrac{E}{R_{(T)}}$ or $\dfrac{V}{R}$

$= \dfrac{2 \times 1 \cdot 5}{4}$

$= 0 \cdot 75$ A

(iii) $P = I^2 R$

$= 0 \cdot 75^2 \times 3 \cdot 6$

$= 2 \cdot 0$ W

(b) Power output is less because:

$P = I^2 R$

Current is less

R (load) is constant

or

$P = \dfrac{V^2}{R}$

t.p.d./V is less

R (load) is constant

or

$P = IV$

Current is less

t.p.d./V is also less

25. (a) Quantity of charge stored per volt

or

Coulombs per volt

or

ratio of charge to p.d./voltage

(b) (i) $3 \cdot 4$ V

(ii) $R = V/I$

$= 3 \cdot 4/0 \cdot 0016$

$= 2125 \ \Omega$

(iii) $V = 12$ V from diagram

$$E = \frac{1}{2} C V^2$$

$$10 \cdot 8 \times 10^{-3} = \frac{1}{2} \times C \times 12^2$$

$$C = 0 \cdot 00015 \text{ F}$$

(c) Time is less as
- Circuit resistance is less
- Current/rate of flow of charge is greater

26. (a) $\dfrac{R_1}{R_2} = \dfrac{R_3}{R_4}$ so $\dfrac{R_{ldr}}{1 \cdot 2} = \dfrac{6}{4}$ so $R_{ldr} = 1 \cdot 8$ (kΩ)

From graph, irradiance $= 0 \cdot 48$ W m^{-2}

(b) (i) $\dfrac{2 \cdot 0}{2 \cdot 0 + 1 \cdot 2} \times 12 = 7 \cdot 5$ V

$$Or \; I = \frac{V}{R_t} = \frac{12}{(1200 + 2000)} = 0 \cdot 00375 \text{ A}$$

$$V = I \times R_{ldr}$$
$$= 0 \cdot 00375 \times 2000$$
$$= 7 \cdot 5 \text{ V}$$

(ii) $V_O = (V_2 - V_1) \dfrac{R_f}{R_1}$

$$= (7 \cdot 2 - 7 \cdot 5) \times \frac{140}{20}$$

$$= -2 \cdot 1 \text{ V}$$

(iii) Yellow LED is lit
because it is **forward biased**.

27. (a) (i) $\dfrac{\sin \theta_a}{\sin \theta_g} = n$

$$\frac{\sin 28}{\sin \theta_g} = 1 \cdot 61$$

$$\theta_g = 17°$$

(ii) $\lambda_{(air)} = \dfrac{v_{(air)}}{f}$

$$= \frac{3 \times 10^8}{4 \cdot 8 \times 10^{14}}$$

$$\lambda_g = \frac{\lambda_{(air)}}{n}$$

$$= \frac{6 \cdot 25 \times 10^{-7}}{1 \cdot 61}$$

$$= 3 \cdot 88 \times 10^{-7} \text{ m}$$

(b) Ray will pass through point X because refractive index for blue light $>$ refractive index for red light **or** blue light refracted more.

28. (a) Power $= 40/20 = 2$ mW
$$P = I \times A$$
$$2 \times 10^{-3} = I \times 8 \times 10^{-5}$$
$$I = 25 \text{ W m}^{-2}$$

(b) For point source, $I_1 \times d_1^2 = I_2 \times d_2^2$
or

$I_1 \times d_1^2 = 1 \cdot 1 \times 0 \cdot 5^2 = 0 \cdot 28$
$I_2 \times d_2^2 = 0 \cdot 8 \times 0 \cdot 7^2 = 0 \cdot 39$
$I_3 \times d_3^2 = 0 \cdot 6 \times 0 \cdot 9^2 = 0 \cdot 49$

Values are not equal - not a point source.

29. (a) *Energy/E/hf*

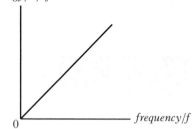

(b) $E = hf$
$$= 6 \cdot 63 \times 10^{-34} \times 6 \cdot 1 \times 10^{14}$$
Photon energy $=$ WF $+ E_k$
WF $= 4 \cdot 044 \times 10^{-19} - 6 \times 10^{-20}$
$$= 3 \cdot 44 \times 10^{-19} \text{ J}$$

(c) Each photon still has same amount of energy.

30. (a) (i) 12 000 decays per second.
(ii) aluminium – 2 half values
lead – 3 half values
800 -- 400 -- 200 -- 100 -- 50 -- 25
count rate = 25 counts per second.

(b) (i) 0·03 μSv

(ii) $\dfrac{60}{0 \cdot 03} = 2000$

PHYSICS HIGHER 2009

SECTION A

1. B	**2.** C	**3.** B	**4.** D
5. C	**6.** A	**7.** C	**8.** D
9. C	**10.** B	**11.** C	**12.** A
13. C	**14.** D	**15.** E	**16.** B
17. D	**18.** E	**19.** A	**20.** E

SECTION B

21. (a) (i) $u_h = 6 \cdot 5 \cos 50° = 4 \cdot 2 \text{ m s}^{-1}$

(ii) $u_v = 6 \cdot 5 \sin 50° = 5 \cdot 0 \text{ m s}^{-1}$

(b) $t = \dfrac{s}{v}$

$= \dfrac{2 \cdot 9}{4 \cdot 2}$

$= 0 \cdot 69 \text{ (s)}$

(c) $s = ut + \frac{1}{2} at^2$

$= 5 \times 0 \cdot 69 + \frac{1}{2} \times -9 \cdot 8 \times (0 \cdot 69)^2$

$= 1 \cdot 1 \text{ (m)}$

so height $h = 2 \cdot 3 + 1 \cdot 1 = 3 \cdot 4 \text{ m}$

(d) Ball would **not** land in basket

(initial) vertical speed would increase

So ball is higher than the basket when it has travelled 2·9 m horizontally

or

So ball has travelled further horizontally when it is at the same height as the basket

22. (a) (i) (A)

$\text{mean} = \dfrac{248 + 259 + 251 + 263 + 254}{5}$

$= 255 \text{ μs}$

(B) $\text{uncertainty} = \dfrac{263 - 248}{5}$

$= (\pm) 3 \text{ μs}$

(ii) (mean contact time $= 255 \pm 3$ μs)

max value $= 258$ μs

club does **not** meet standard

(b) (i) $F = \dfrac{mv - mu}{t}$

$= \dfrac{4 \cdot 5 \times 10^{-2} \times (50 - 0)}{450 \times 10^{-6}}$

$= 5000 \text{ N}$

(ii) Impulse on the ball is greater

or

$\underline{\Delta}$mv is greater

So speed increased

23. (a) (i) $P \times V = 2000 \ 1995 \ 2002 \ 2001$

all 4 values needed

$P \times V = \text{constant}$

or $P \times V = 2000$

or $P_1 V_1 = P_2 V_2$

or $P = k/V$

(ii) Gas <u>molecules</u> <u>collide</u> with <u>walls</u> of container more often so (average) force increases

pressure increases

(b) (i) pressure due to <u>water</u>

$P = \rho g h$

$= 1020 \times 9 \cdot 8 \times 12$

$= 120000 \text{ (Pa)}$

$\text{Total pressure} = 120000 + 1 \cdot 01 \times 10^5$

$= 2 \cdot 21 \times 10^5 \text{ (Pa)}$

(ii) $P_1 V_1 = P_2 V_2$

$1 \cdot 01 \times 10^5 \times 1 \cdot 50 \times 10^{-3} = 2 \cdot 21 \times 10^5 \times V_2$

$V_2 = 6 \cdot 86 \times 10^{-4} \text{ m}^3$

(c) pressure decreases as $P = \rho g h$

volume of air in lungs will increase

(or pressure difference increases)

so <u>lungs</u> may become damaged

24. (a) (i) $V_{tpd} = IR$

$= 1 \cdot 5 \times 3$

$= 4 \cdot 5 \text{ (V)}$

$\text{lost volts} = E - V_{tpd}$

$= 6 \cdot 0 - 4 \cdot 5$

$= 1 \cdot 5 \text{ V}$

(ii) $r = \dfrac{\text{lost volts}}{I}$

$= \dfrac{1 \cdot 5}{3 \cdot 0}$

$= 0 \cdot 5 \ \Omega$

or

$r = \dfrac{E}{I}$

$= \dfrac{6 \cdot 0}{12}$

$= 0 \cdot 5 \ \Omega$

or

$E = IR + Ir$

$6 \cdot 0 = (3 \times 1 \cdot 5) + (3 \times r)$

$r = 0 \cdot 5 \ \Omega$

(b) current decreases

so lost volts $(V = Ir)$ decreases

25. (a) (i) $V_P = 3 \times 0 \cdot 5 = 1 \cdot 5 \text{ mV}$

(ii) $f = \dfrac{1}{T}$

$= \dfrac{1}{4 \times 10^{-3}}$

$= 250 \text{ Hz}$

(b) (i) inverting (mode)

(ii) $V_{rms} = \dfrac{V_{peak}}{\sqrt{2}}$

$= \dfrac{6 \cdot 2 \times 10^{-3}}{\sqrt{2}}$

$= 4 \cdot 38 \times 10^{-3} \ (V)$

$\dfrac{V_O}{V_I} = -\dfrac{R_f}{R_l}$

$\dfrac{V_O}{4 \cdot 38 \times 10^{-3}} = -\dfrac{10 \times 10^6}{5 \times 10^3}$

$V_O = (-) \ 8 \cdot 8 \ V$

(iii) trace will be "clipped"/flattened (at $\pm$ 9 V) or <u>almost</u> square wave
max output voltage will be $\pm$ 9 V/V_s
or op-amp saturates
or saturation occurs

26. (a) (Current)

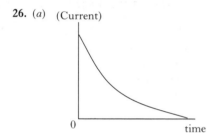

(b) $V_R = I R$

$= 5 \times 10^{-3} \times 500$

$= 2 \cdot 5 \ (V)$

$V_C = 12 - 2 \cdot 5$

$= 9 \cdot 5 \ V$

(c) $E = \dfrac{1}{2} C V^2$

$= 0 \cdot 5 \times 47 \times 10^{-6} \times 12^2$

$= 3 \cdot 4 \times 10^{-3} \ J$

(d) Max energy the same/ 'no effect'
Values of "C" <u>and</u> "V" are same as before

27. (a) waves <u>meet</u> out of phase
or crest meets trough
or path difference $= (n + \frac{1}{2}) \lambda$

(b) $\lambda_{blue \ light}$ is shorter (than $\lambda_{red \ light}$)
and $n \lambda = d \sin\theta$
or
$\sin\theta = n \lambda/d$

(c) $n \lambda = d \sin\theta$
$2 \times 4 \cdot 73 \times 10^{-7} = 2 \cdot 00 \times 10^{-6} \sin\theta$
$\theta = 28 \cdot 2°$

28. (a) (i) $E_3 \rightrightarrows E_0$

$(\Delta)E \ \alpha \ f$ **or** $E = h f$

$f \alpha \dfrac{1}{\lambda}$ **or** $v = f\lambda$

(ii) $(\Delta)E = h f$ or $W_2 - W_1 = h f$

$- 5 \cdot 2 \times 10^{-19} - (-9 \cdot 0 \times 10^{-19}) = 6 \cdot 63 \times 10^{-34} \times f$

$f = 5 \cdot 7 \times 10^{14} \ Hz$

(b) $\lambda_a = \left(\dfrac{v}{f}\right) = \dfrac{3 \times 10^8}{4 \cdot 6 \times 10^{14}}$

$= 6 \cdot 5 \times 10^{-7} \ (m)$

$\dfrac{\lambda_a}{\lambda_g} = \dfrac{\sin\theta_a}{\sin\theta_g}$

$\dfrac{6 \cdot 5 \times 10^{-7}}{\lambda_g} = \dfrac{\sin 53°}{\sin 30°}$

$\lambda_g = 4 \cdot 1 \times 10^{-7} \ m$

29. (a) (i) $E_k = h f - h f_0$

$= 5 \cdot 23 \times 10^{-19} - 2 \cdot 56 \times 10^{-19}$

$= 2 \cdot 67 \times 10^{-19} \ J$

(ii) $E_k = \dfrac{1}{2} m v^2$

$2 \cdot 67 \times 10^{-19} = \dfrac{1}{2} \times 9 \cdot 11 \times 10^{-31} \times v^2$

$v = 7 \cdot 66 \times 10^5 \ ms^{-1}$

(b) No change (to maximum speed)/no effect
Energy/frequency of photons does not change
or
Energy an electron receives is the same

30. (a) (i) $r = 95$

$s = 7$

(ii) Total mass of reactants
$>$ total mass of products
or
(there is a) loss of mass

(iii) Total mass before
$= 390 \cdot 173 \times 10^{-27} + 1 \cdot 675 \times 10^{-27}$
$= 3 \cdot 91848 \times 10^{-25} \ (kg)$

Total mass after
$= 230 \cdot 584 \times 10^{-27} + 157 \cdot 544 \times 10^{-27} +$
$(2 \times 1 \cdot 675 \times 10^{-27})$
$= 3 \cdot 91478 \times 10^{-25} \ (kg)$

$\Delta m = 3 \cdot 91848 \times 10^{-25} - 3 \cdot 91478 \times 10^{-25}$
$= 3 \cdot 7 \times 10^{-28} \ (kg)$

$E = mc^2$
$= 3 \cdot 7 \times 10^{-28} \times (3 \times 10^8)^2$
$= 3 \cdot 3 \times 10^{-11} \ J$

(b) (i) 12 mm

(ii) $200 \rightarrow 100 \rightarrow 50$
2 half-value thicknesses
$= 2 \times 12 = 24 \ mm$

PHYSICS HIGHER
2010

SECTION A

1. E	2. E	3. D	4. A
5. D	6. B	7. C	8. B
9. A	10. D	11. D	12. C
13. E	14. A	15. B	16. D
17. E	18. D	19. B	20. C

See also the extra note sheets.

SECTION B

21. (a) (i) 47 km
    $156^{(o)}$
    or 24° east of south
    or 66° south of east

    (ii) $v = s/t$
    $v = (47100$ **or** $47000)/900$
    $v = 52 \cdot 3$ **or** $52 \cdot 2$ m s^{-1}
    [**or** 188 km h^{-1}]
    at $156^{(o)}$

    (b) (i) Lift = mg **or** lift = weight
    or forces balanced
    $W = 1 \cdot 21 \times 10^4 \times 9 \cdot 8$
    $W = 119$ kN

    (ii) Weight is less
    There is a resultant force upwards **or** unbalanced force upwards **or** net force upwards

    Upward Acceleration

    OR

    The helicopter accelerates upwards
    weight is less
    there is a net upward force

22. (a) (i) The total momentum before (a collision) equals the total momentum after (the collision)

    'In the absence of external forces'
    or 'in an isolated system'

    (ii) $m_A u_A + m_B u_B = m_A v_A + m_B v_B$
    $(0 \cdot 22 \times 0 \cdot 25) + 0 \cdot 16u = (0 \cdot 38 \times 0 \cdot 2)$
    $0 \cdot 055 + 0 \cdot 16u = 0 \cdot 076$
    $u = 0.13$ m s^{-1}

    (b) The find velocity is less
    because the (total initial) momentum is less,
    the mass is constant
    and v = momentum/mass

23. (a) (i) $v^2 = u^2 + 2as$
    $v^2 = 0^2 + 2 \times 9 \cdot 8 \times 2$
    $v = \underline{6 \cdot 3}$ (m s^{-1})
    or
    $(m)gh = \frac{1}{2}(m)v^2$
    $v = \sqrt{(2 \times 9 \cdot 8 \times 2)}$
    $v = \underline{6 \cdot 3}$ (m s^{-1})

    (ii) $(\Delta p) = m (v - u)$
    $= 40 (-5.7 - 6.3)$
    $= -480$ kg m s^{-1}

    or

    $(\Delta p) = m (v - u)$
    $= 40 (5.7 - (-6.3))$
    $= 480$ kg m s^{-1}

    (iii) $F = \Delta p / t$
    $F = (-)480/0 \cdot 5$
    $F = (-)960$ N

    (b) Weight/downwards force is constant
    vertical component(s) balances weight

    as angle increases tension must increase
    because T = ½ W/ cos θ

24. (a) (i) $0 \cdot 51$ s

    (ii) Random uncertainty = {max – min} / no.
    $= \{0 \cdot 55 - 0 \cdot 49\}/6$
    $= 0 \cdot 01$ s

    (b) (i) $Q = CV$
    $Q = 1 \cdot 6 \times 10^{-3} \times 4 \cdot 5$
    $Q = 7 \cdot 2 \times 10^{-3}$ C

    (ii) $(I = V/R$
    $I = 4 \cdot 5/18000$
    $I = 0 \cdot 25$ mA)

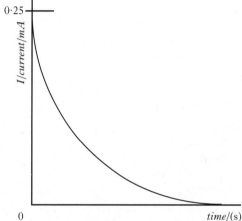

25. (a) (i) $R_1/R_2 = R_3/R_4$
    $R_1 = 6000 \times 800/4000$
    $R_1 = 1200$ Ω

    (ii) $V_P = 4 \cdot 0$ V
    $V_Q = 4 \cdot 8$ V
    Voltmeter reading = $0 \cdot 8$ V

    (b) $V_o = (V_2 - V_1) (R_f/R_1)$
    $V_o = (3 \cdot 2 - 3 \cdot 0) (2 \cdot 0 \times 10^6/20 \times 10^3)$
    $V_o = 20$ (V)

    (But, due to saturation, the actual output voltage is)
    10 to 12 V

26. (a) (i) $V_p = 2 \cdot 0$ V

    (ii) $f = 1/T$
    $f = 1/0 \cdot 01$
    $f = 100$ Hz

    (b) Stays the same/constant/no change/nothing

    (c) increases/doubles

    (d) The capacitor will be damaged
    The peak voltage from this power supply is greater than 16 V because $V_P = \sqrt{2} \times 15 = 21.2$ V

27. (a) $S_2 P - S_1 P = (n + \frac{1}{2}) \lambda$
    $0 \cdot 34 = \lambda/2$
    $\lambda = 0 \cdot 68$ m

    or

    path difference = ½ λ
    path difference = $0 \cdot 34$ m
    $\lambda = 0 \cdot 68$ m

(b) The amplitude increases **or** is greater

because <u>destructive</u> interference is no longer taking place

28. (a) (i) $P = F/A$

$F = 4\cdot6 \times 10^5 \times 3\cdot00 \times 10^{-2}$

$F = 13800$ N

(ii) $P_1V_1 = P_2V_2$

$4\cdot6 \times 10^5 \times 1\cdot6 \times 10^{-3} = 1\cdot0 \times 10^5 \times V_2$

$V_2 = 7\cdot36 \times 10^{-3}$ (m^3)

V of water $= (7\cdot36 - 1\cdot6) \times 10^{-3}$

$= 5\cdot76 \times 10^{-3}$ m^3

(b) (i) Stays the same/constant/nothing/no change

(ii) $n = \sin \theta_1/\sin \theta_2$

$n = \sin 60/\sin 41$

$n = 1\cdot32$

(iii) $\sin \theta_C = 1/n$

$\sin \theta_C = 1/1\cdot32$

$\theta_C = 49°$

(iv) The critical angle is less

because the refractive index is larger

29. (a) Very small area/diameter/radius (of beam)

$I = P/A$ **or** High irradiance

(b) $E = hf$

$E = 6\cdot63 \times 10^{-34} \times 4\cdot74 \times 10^{14}$

$E = 3\cdot14 \times 10^{-19}$ J

(c) Frequency/wavelength/energy

Direction

Speed

Phase/coherent

Velocity

(d) $\lambda = v/f = 3 \times 10^8/4\cdot74 \times 10^{14} = 633$ (nm)

$n\lambda = d \sin \theta$

$d = (2 \times 633 \times 10^{-9})/\sin 30$

$d = 2\cdot5 \times 10^{-6}$ m

30. (a) 146

(b) (i) $r = 93$

$s = 237$

(ii) $T =$ Neptunium (**or** Np)

(c) $N = At$

$N = 30 \times 10^3 \times 60$

$N = 1\cdot8 \times 10^6$

(d) $I = V/R$

$I = 5/16$

$I = 0\cdot3125$ (A)

$E = I(R + r)$

$9 = 0\cdot3125\ (R + 2)$

$9 = 0\cdot3125R + 0\cdot625$

$8\cdot375 = 0\cdot3125R$

$R = 26\cdot8\ \Omega$

$R = 26\cdot8 - 16 = 10\cdot8 = 11\ \Omega$

or

$I = V/R$

$I = 5/16$

$I = 0\cdot3125$ (A)

$V_{lost} = Ir = 2 \times 0\cdot3125 = 0\cdot625$ (V)

$V_{resistor} = 9 - (5 + 0\cdot625) = 3\cdot375$ (V)

$R = V/I$

$R = 3\cdot375/0\cdot3125$

$R = 10\cdot8 = 11\Omega$

or

$I = V/R$

$= 5/16$

$= 0\cdot3125$ (A)

$R_T = E/I$

$= 9/0\cdot3125$

$= 28\cdot8\Omega$

$R = R_T - 18 = 28\cdot8 - 18$

$= 10\cdot8 = 11\Omega$

PHYSICS HIGHER 2011

SECTION A

1. C	2. E	3. C	4. A
5. C	6. D	7. A	8. E
9. C	10. B	11. B	12. A
13. D	14. E	15. D	16. B
17. D	18. A	19. B	20. A

SECTION B

21. (a) (i) $v^2 = u^2 + 2as$

$0 = 7^2 + 2 \times (-9\cdot8) \times s$

$s = \mathbf{2\cdot5}\,\text{m}$

or

$v = u + at$

$0 = 7 + (-9\cdot8)\,t$

$t = 0\cdot71\,\text{s}$

$s = ut + \frac{1}{2}\,a\,t^2$

$= 7 \times 0\cdot71 + \frac{1}{2}\,(-9\cdot8)\,(0\cdot71)^2$

$= \mathbf{2\cdot5}\,\text{m}$

(ii) $v = u + at$

$0 = 7 + (-9\cdot8) \times t$

$t = \mathbf{0\cdot71}\,\text{s}$

or

$s = \left(\dfrac{u + v}{2}\right)t$

$2\cdot5 = \left(\dfrac{7 + 0}{2}\right) \times t$

$t = \mathbf{0\cdot71}\,\text{s}$

(b) (i) $1\cdot5\,\text{m s}^{-1}$ to the **right**

(ii) Statement Z

Horizontal speed of ball remains constant and equal to (horizontal) speed of trolley

or

Horizontal speed of the ball remains constant at $1\cdot5\,\text{m s}^{-1}$

22. (a) (i) *Impulse = Area under F–t graph*

$= \frac{1}{2}\,6\cdot4 \times 0\cdot25$

$= \mathbf{0\cdot80}\,\text{kg m s}^{-1}$

(ii) $\mathbf{0\cdot80}\,\text{kg m s}^{-1}$

in the negative direction **or** to the left

(iii) *(Impulse = Change in momentum)*

$F \times t = mv - mu$

$-0\cdot80 = m\,(-0\cdot45 - 0\cdot48)$

$m = \mathbf{0\cdot86}\,\text{kg}$

(b)

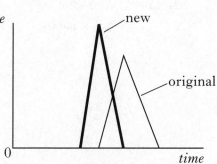

23. (a) (i) $m = 111\cdot49 - 111\cdot26$

$= 0\cdot23\,\text{g}$

$\rho = m/V$

$= 0\cdot23 \times 10^{-3}/2\cdot0 \times 10^{-4}$

$= \mathbf{1\cdot15}\,\text{kg m}^{-3}$

(ii) Not all the air will be evacuated from jar

or

It is impossible to get a (perfect) vacuum

or

Some air has leaked back in

(b) (i) $P_1 V_1 = P_2 V_2$

$1\cdot01 \times 10^5 \times 200 = P_2 \times 250$

$P_2 = \mathbf{8\cdot1 \times 10^4}\,\text{Pa}$

(ii) <u>Particles collide</u> with <u>walls</u> of jar

So when air is removed, number of collisions on walls of jar is less frequent/less often and average force (on walls) decreases, and pressure on walls of jar decreases

24. (a) (i) 10 joules of energy are given to each coulomb (of charge) passing through the supply

(ii) $I = \dfrac{E}{(R + r)}$

$1\cdot25 = \dfrac{10}{(6 + r)}$

$r = \mathbf{2\cdot0}\,\Omega$

or

$r = \dfrac{lost\ volts}{I}$

$= \dfrac{10 - 7\cdot5}{1\cdot25}$

$= \mathbf{2\cdot0}\,\Omega$

or

$\dfrac{R_1}{R_2} = \dfrac{V_1}{V_2}$

$\dfrac{r}{6\cdot0} = \dfrac{2\cdot5}{7\cdot5}$

$r = \mathbf{2\cdot0}\,\Omega$

(b) (i) (Total) resistance decreases

(circuit) current increases

lost volts increases

(ii) Parallel resistance $= R = V/I$

$= 6\cdot0/2\cdot0$

$= 3\cdot0\,\Omega$

$1/R_T = 1/R_1 + 1/R_2$

$1/3 = 1/6 + 1/R$

$R = \mathbf{6\cdot0}\,\Omega$

or

Total resistance $= E/I$

$= 10/2\cdot0$

$= 5\cdot0\,\Omega$

Resistance of parallel network $= 5 - 2$

$= 3\,\Omega$

$R_T = \dfrac{Product}{Sum}$

$3 = \dfrac{6 \times R}{6 + R}$

$R = \mathbf{6\cdot0}\,\Omega$

25. (a) 200 µC of charge increases voltage across plates by 1 volt

    or

    200 µC per volt

    or

    One volt across the plates of the capacitor causes 200 µC of charge to be stored

  (b) (i) $I = E/R$

        $= 12/1400$

        $= \textbf{0·0086}$ A

          (8·6 mA)

    (ii) $E = \frac{1}{2}\,CV^2$

      initial stored energy $= \frac{1}{2} \times (200 \times 10^{-6}) \times 12^2$

               $= 0·0144$ J

      final stored energy $= \frac{1}{2}\,(200 \times 10^{-6}) \times 4^2$

               $= 0·0016$ J

      Difference $= 0·0144 - 0·0016$

      decrease in stored energy $= \textbf{0·0128}$ **J**

  (c) (i) 0·30 s

    (ii) $s = ut + \frac{1}{2}\,a\,t^2$

      $0·80 = 1·5 \times 0·3 + \frac{1}{2} \times a \times (0·3)^2$

         $a = \textbf{7·8}$ **m s**$^{-2}$

    (iii) Percentage (fractional) uncertainty in (measuring) <u>distance</u> will be smaller

      or

      Percentage (fractional) uncertainty in (measuring) <u>time</u> will be smaller

26. (a) (i) Inverting

    (ii) $V_o = -\dfrac{R_f}{R_1} \times V_1$

       $12 = \dfrac{-80}{10} \times V_1$

       $V_1 = -\textbf{1·5}$ **V**

    (iii) Output cannot be greater than (approx 85% of) the supply voltage

      or

      Saturation <u>of the amplifier</u> has been reached

  (b) V_o/V

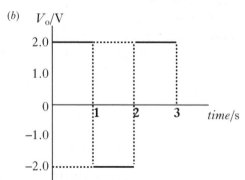

27. (a) (i) $n = \dfrac{sin\,\theta_1}{sin\,\theta_2}$

      $1·66 = \dfrac{sin\,40}{sin\,\theta}$

        $\theta = \textbf{22·8°}$

    (ii) (A) $sin\,\theta_C = 1/n$

                $= 1/1·66$

          $\theta_C = \textbf{37·0°}$

      (B) **74°**

(b) No

   or

   it is totally internally reflected $\Big\}$

  and

   n depends on frequency

   or

   $n_{blue} > n_{red}$

   or

   blue refracts more than red $\Big\}$

  and

   (critical angle) $_{blue}$ < (critical angle) $_{red}$ $\Big\}$

   or

   the angle of incidence has increased $\Big\}$

  and

   angle of incidence of blue light on face PQ is greater than the critical angle

28. (a) Light travels as waves

    or

    Energy in light is carried as a wave

    or

    Light is a wave

  (b) (i) $d\,sin\,\theta = n\lambda$

      $5 \times 10^{-6} \times sin\,11 = 2 \times \lambda$

                      $\lambda = \textbf{480}$ **nm**

    (ii) Spacing of maxima increases

      λ in liquid increases

      (as n decreases)

      $sin\,\theta = n\lambda/d$

      θ increases

29. (a) (i) $f = \dfrac{c}{\lambda}$

          $= \dfrac{3·00 \times 10^8}{525 \times 10^{-9}}$

          $= 5·71 \times 10^{14}$ Hz

     $E = hf$

          $= 6·63 \times 10^{-34} \times 5·71 \times 10^{14}$

          $= \textbf{3·79} \times \textbf{10}^{-19}$ **J**

    (ii) $\left(\begin{array}{l} E_k = hf - hf_o \\ = 3·79 \times 10^{-19} - 2·24 \times 10^{-19} \end{array}\right)$

        $= \textbf{1·55} \times \textbf{10}^{-19}$ **J**

  (b) (i) <u>Photons</u> with frequency below f_o do not have enough <u>energy</u> to release electrons

      or

      <u>Photons</u> with frequency below f_o have <u>energy</u> smaller than work function

    (ii) Work function $= hf_o$ (**or** $E = hf_o$)

      $2·24 \times 10^{-19} = (6·63 \times 10^{-34}) \times f_o$

           $f_o = \textbf{3·38} \times \textbf{10}^{14}$ **Hz**

30. (*a*) (i) (Nuclear) Fusion

(ii) Total mass before

$= 3 \cdot 342 \times 10^{-27} + 5 \cdot 005 \times 10^{-27}$

$= 8 \cdot 347 \times 10^{-27}$ (kg)

Total mass after

$= 6 \cdot 642 \times 10^{-27} + 1 \cdot 675 \times 10^{-27}$

$= 8 \cdot 317 \times 10^{-27}$ (kg)

Loss in mass $= 0 \cdot 030 \times 10^{-27}$ (kg)

Energy released $= mc^2$

$= 0 \cdot 030 \times 10^{-27} \times (3 \cdot 00 \times 10^8)^2$

$= \mathbf{2 \cdot 7 \times 10^{-12}}$ **J**

(*b*) (i) Energy absorbed

$= -1 \cdot 360 \times 10^{-19} - (-5 \cdot 424 \times 10^{-19})$

$= 4 \cdot 064 \times 10^{-19}$(J)

$E = hf$

$4 \cdot 064 \times 10^{-19} = 6 \cdot 63 \times 10^{-34} \times f$

$f = 6 \cdot 13 \times 10^{14}$ (Hz)

$\lambda = \dfrac{c}{f}$

$= \dfrac{3 \cdot 00 \times 10^8}{6 \cdot 13 \times 10^{14}}$

$= \mathbf{489}$ nm

(ii) 'Blue' **or** 'blue-green'

PHYSICS HIGHER
2012

SECTION A

1. E	2. A	3. C	4. C
5. C	6. D	7. B	8. D
9. B	10. C	11. B	12. E
13. D	14. E	15. D	16. A
17. A	18. D	19. B	20. B

SECTION B

21. (*a*) (i)

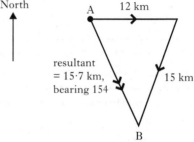

(½) for correct diagram to scale, length and angle

(½) for adding correctly showing resultant direction (arrow needed)

displacement $= 15 \cdot 7 \pm 0 \cdot 3$ km

bearing $= 154 \pm 2$ (26° E of S)

(64° S of E)

(ii) $v = \dfrac{s}{t}$

$= \dfrac{15 \cdot 7}{1 \cdot 25}$

$= 12 \cdot 6$ km h^{-1} at 154

(*b*) (i) 15·7 km on a bearing of 154

(ii) $t = \dfrac{d}{v}$

$= \dfrac{33}{22}$

$= 1 \cdot 5$ hours

$v = \dfrac{s}{t}$

$= \dfrac{15 \cdot 7}{1 \cdot 5}$

$= 10 \cdot 5$ km h^{-1} on a bearing of 154

22. (*a*) (i) $d = vt$

$= 20 \times 3 \cdot 06$

$= 61 \cdot 2$ m

(ii) $v^2 = u^2 + 2as$

$0 = 15^2 + 2 \times -9 \cdot 8 \times s$

$s = 11 \cdot 5$ m

(11·48)

(*b*) More likely, because:

horizontal velocity will decrease

range will decrease

time in air will decrease

height reached will decrease

23. (a) (i) $E_w = Q V$
$$= 1 \cdot 6 \times 10^{-19} \times 1220$$
$$= 1 \cdot 95 \times 10^{-16} \text{ J}$$

(ii) work done $= \frac{1}{2} m v^2$
$$= \frac{1}{2} \times 2 \cdot 18 \times 10^{-25} \times v^2$$
$$v = 4 \cdot 23 \times 10^4 \text{ m s}^{-1}$$

(b) $Ft = \Delta mv$
$$0 \cdot 07 \times 60 = 750 \times \Delta v$$
$$\Delta v = 5 \cdot 6 \times 10^{-3} \text{ m s}^{-1}$$

(c) Force from Xenon engine greater

Change in momentum of the Xenon ions would be greater (than Krypton ions)

Impulse from Xenon ions would be greater

24. (a)

P/T	347	347	346	348	348

Pressure and temperature (in K) are directly proportional

(b) As temperature increases, E_k of gas molecules/particles increases (**or** molecules travel faster) and hit/collide with the walls of the container more often/frequently

with greater force

pressure increases

(c) To ensure all the gas in the flask is heated evenly

or all the gas is at the same temperature

25. (a) (i) $I = \dfrac{E}{(R + r)}$
$$= \dfrac{12}{(6 + 2)}$$
$$= 1 \cdot 5 \text{ A}$$

(ii) $V = Ir$
$$= 1 \cdot 5 \times 2$$
$$= 3 \cdot 0 \text{ V}$$

(iii) $P = I^2 R$
$$= (1 \cdot 5)^2 \times 6$$
$$= 13 \cdot 5 \text{ W}$$
or
$P = V^2/R$
$$= 9^2/6$$
$$= 13 \cdot 5 \text{ W}$$
or
$P = IV$
$$= 1 \cdot 5 \times 9$$
$$= 13 \cdot 5 \text{ W}$$

(b) $P = I^2 R$

(Circuit) current increases
Total or circuit resistance decreases
Internal resistance less
or
$P = V^2/R$
Voltage across lamp increases
Lost volts decreases
Internal resistance less

26. (a)

(b) $R = V/I$
$$= \dfrac{12}{2 \times 10^{-3}}$$
$$= 6000 \ \Omega$$
$$(6 \cdot 0 \text{ k}\Omega)$$

(c) (i) Initial current only depends on the values of the e.m.f. of the supply <u>and</u> resistor R which do not change.

(ii) Smaller
Capacitor takes less time to discharge

27. (a) Resistance of fabric = 40 Ω
$$\dfrac{R_1}{R_2} = \dfrac{R_3}{R_4}$$
$$\dfrac{R_V}{40} = \dfrac{240}{80}$$
$$R_V = 120 \ \Omega$$

(b) (i) Differential Mode

(ii) Gain $= \dfrac{R_f}{R_1}$
$$= \dfrac{560}{100}$$
$$= 5 \cdot 6$$

(iii) (A) Gain $= \dfrac{V_{out}}{V_{in}}$
$$5 \cdot 6 = \dfrac{10 \cdot 8}{V_{in}}$$
$$V_{in} = 1 \cdot 93 \text{ V}$$

(B) Potential at X $= 2 \cdot 25 + 1 \cdot 93$
$$= 4 \cdot 18 \text{ V}$$
$$\dfrac{R_1}{R_2} = \dfrac{V_1}{V_2}$$
$$\dfrac{R_1}{120} = \dfrac{4 \cdot 18}{4 \cdot 82}$$
$$R_1 = 104 \ \Omega$$
Length of fabric = 66 mm

28. (a) $n = \dfrac{\sin \theta_1}{\sin \theta_2}$
$$1 \cdot 33 = \dfrac{\sin X}{\sin 36}$$
$$X = 51°$$

(b) (i) Angle of refraction is 90°
or
Refracted ray makes an angle of 90° with normal
or
Refracted ray is along surface of water

(ii) $\sin \theta_C = 1/n$
$\qquad = 1/1\cdot33$
$\qquad \theta_C = \mathbf{49°}$

(c)

Totally internally reflected ray shown (angles should be equal)

29. (a) $\qquad d\sin\theta = n\lambda$
$d \times \sin 35\cdot3 = 3 \times 633 \times 10^{-9}$
$\qquad d = \mathbf{3\cdot29 \times 10^{-6}\ m}$

(b) Number of lines per metre $= \dfrac{1}{3\cdot29 \times 10^{-6}}$

$\qquad\qquad = \mathbf{3\cdot04 \times 10^{5}}$

(c) $\qquad$ Difference $= (3\cdot04 - 3\cdot00) \times 10^5$

$\qquad\qquad = 0\cdot04 \times 10^5$

Percentage difference $= \dfrac{0\cdot04 \times 10^5}{3\cdot00 \times 10^5} \times 100$

$\qquad\qquad = 1\cdot33\%$

Technician's value <u>does</u> agree

30. (a) Decreases

(b) (i) Photoconductive mode

(ii) Current increases

more photons of light arrive at the junction per second

more free charge carriers produced per second

(c) $\qquad I_1 d_1^{\,2} = I_2 d_2^{\,2}$
$3\cdot0 \times 10^{-6} \times 1\cdot2^2 = I_2 \times 0\cdot8^2$
$\qquad\qquad I_2 = 6\cdot75\ \mu A$

31. (a) $\qquad D = \dfrac{E}{m}$

$500 \times 10^{-6} = \dfrac{E}{0\cdot04}$

$\qquad E = 2\cdot0 \times 10^{-5}\ J$

(b) $\qquad \overset{\bullet}{H} = \dfrac{H}{t}$

$5\cdot0 \times 10^{-3} = \dfrac{H}{2}$

$\qquad H = 0\cdot01\ (Sv)$

$\qquad H = Dw_R$

$0\cdot01 = 500 \times 10^{-6} \times w_R$

$\qquad w_R = 20$

alpha radiation

Hey! I've done it

BrightRED
PUBLISHING

© 2012 SQA/Bright Red Publishing Ltd, All Rights Reserved
Published by Bright Red Publishing Ltd, 6 Stafford Street, Edinburgh, EH3 7AU
Tel: 0131 220 5804, Fax: 0131 220 6710, enquiries: sales@brightredpublishing.co.uk,
www.brightredpublishing.co.uk

Official SQA answers to 978-1-84948-297-4
2008-2012